Essential Astrophysics

Essential Astrophysics
Interstellar Medium to Stellar Remnants

Shantanu Basu and Pranav Sharma

CRC Press
Taylor & Francis Group
Boca Raton London New York

CRC Press is an imprint of the
Taylor & Francis Group, an **informa** business

First edition published 2022
by CRC Press
6000 Broken Sound Parkway NW, Suite 300, Boca Raton, FL 33487-2742

and by CRC Press
2 Park Square, Milton Park, Abingdon, Oxon, OX14 4RN

© 2022 Taylor & Francis Group, LLC

CRC Press is an imprint of Taylor & Francis Group, LLC

Library of Congress Cataloging-in-Publication Data

ISBN: 978-0-367-76847-8 (hbk)
ISBN: 978-1-032-10563-5 (pbk)
ISBN: 978-1-003-21594-3 (ebk)

DOI 10.1201/9781003215943

Typeset in Latin Modern font
by KnowledgeWorks Global Ltd.

Contents

List of Figures

Preface

This text grew out of several courses on introductory astrophysics that S. B. taught at The University of Western Ontario, Canada. The idea was always to write up the set of course notes with some additional explanations and diagrams, into a readable compendium. One that a student with a modest background in undergraduate physics and mathematics could use to quickly get up to speed on important results and concepts in the astrophysics of stars, from their birthplaces to their end states. In recent years, we organized several winter schools on astronomy that were geared to undergraduate university students. What we realized is that there is a great enthusiasm amongst students to learn astronomical ideas and derivations, even when those students are not currently in a physics or astronomy program. Many students are enrolled in quantitative subjects, most notably in engineering programs, but they are eager to take the plunge into the science of astronomy.

This book is a humble contribution to the literature of quantitative introductory astronomy texts. It may be your first exciting step into astrophysics. Our goal is to be concise, readable, and intuitive, yet quantitative as needed. There are only a limited number of books pitched at the level suitable for an introductory undergraduate course for STEM subject area students. Many of those books are older and are now hard to obtain. For a very comprehensive coverage, one turns to *An Introduction to Modern Astrophysics*, by Carroll and Ostlie. Another good resource is *Introductory Astronomy and Astrophysics*, by Zeilik and Gregory. For an even older resource that has unbeatable intuitive explanations, one need look no further than *The Physical Universe*, by Shu. Our book is compact and focuses on the formation and life history of stars, starting with the backdrop of the interstellar medium (ISM). An emphasis on the ISM and star formation reflects S. B.'s own research interests and background. We have included recent results from modern astronomical observatories including the *Hubble Space Telescope*, the *Atacama Large Millimeter/Submillimeter Array (ALMA)*, the *Stratospheric Observatory for Infrared Astronomy (SOFIA)*, the *Planck* satellite, the *Kepler* satellite, and the *Laser Interferometric Gravitational Wave Observatory (LIGO)*.

The last decade has seen an unprecedented expansion of access to the internet and e-resources around the world. Since 2010, just India and China together have added about a billion more people who have access to the internet. This is a unique time in human history when a large number of people are becoming globally aware and connected. The implications are enormous, and we hope that we can reach readers in all corners of the globe, on whatever platform they are using.

This book would not have been possible without the contribution of many who helped with proofreading and with figure development. Special thanks to Arjun Kota who worked diligently on many figures throughout the writing process, along with Varnit Shrivastava, and Sawrav Cintury. We also thank Gianfranco Bino and Indrani Das for making some figures. Proofreading by Najeh Jisrawi, Sayantan Auddy, Deepakshi Madaan, Pranav Manangath, Indrani Das, and Mahmoud Sharkawi is greatly appreciated. S. B. is grateful to Telemachos Mouschovias for introducing him to the wonders of the ISM and Star Formation, and to the many inquisitive students he has taught and supervised over the years. P. S. would like to thank the B. M. Birla Science Centre, India, for their encouragement. He also thanks Bruce Ferguson, Iqbal Patni, Ravi Singh, and Harvijay Singh Bahia for their support.

Shantanu Basu
London, Canada

Pranav Sharma
New Delhi, India

March, 2021

CHAPTER 1

Introduction

1.1 OVERVIEW

We live on a planet that was formed in a swirling disk of matter around a star that was born in a stellar cluster birthed from a massive interstellar gas cloud. This book traces a path from interstellar matter to star-disk-planet systems and the evolution of the stars that ultimately leads to stellar remnants such as white dwarfs, neutron stars, and black holes.

This story plays out within the confines of our Galaxy, a swirling disk and halo of stars and gas, shown schematically in Figure 1.1. The gaseous disk contains majestic spiral arm patterns, and there are additional components to the interstellar medium, e.g., a large-scale magnetic field and fast-moving charged particles known as cosmic rays. Indirect evidence also implies a much larger Galactic halo of invisible **dark matter** that contains more mass than all the visible components.

DOI: 10.1201/9781003215943-1

Figure 1.1 An illustration of the visible components of our Galaxy. The Sun is located in an outer part of the Galactic disk that is characterized by majestic spiral arms. There is also a bulge of stars near the center, as well as a large sphere-like halo of stars that includes the dense globular star clusters.

The Sun and its associated solar system occupy the middle suburbs of the Galaxy, orbiting the center at a radius of about 8 kpc. The solar system itself has a disk-like structure, but its plane is tilted relative to that of the Galaxy by a remarkable 60.2°. Nature refuses to give us the simplicity of alignment! The Earth's equator is also tilted relative to the ecliptic, by 23.4°. Viewed from the Earth, we see all stars, planets, galaxies, gas clouds, etc. occupying a fixed or changing position, as the case may be, on an imaginary **celestial sphere** that envelops the Earth. Positions are measured on the celestial sphere using coordinates of declination (δ) and right ascension (α) that are counterparts to latitude and longitude, respectively, that are used on the Earth's surface. The **celestial equator** is an imaginary circle that is the projection of the Earth's equator onto the celestial sphere. Figure 1.2 illustrates the concept of the celestial sphere.

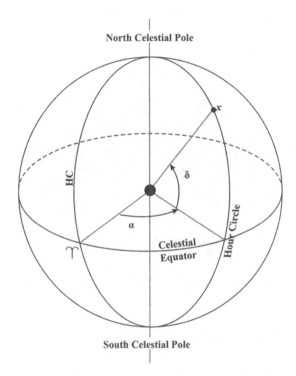

Figure 1.2 An illustration of the celestial sphere. The north celestial pole and south celestial pole are located at the extensions of the Earth's north and south pole, respectively. The celestial equator is the great circle that is in the plane defined by the Earth's equator. An hour circle is any great circle that runs through both the north and south celestial poles. The vernal equinox (♈) is located at the intersection of the ecliptic and the celestial equator that the Sun crosses once each year while moving into the northern celestial hemisphere. The position of a celestial object (approximately fixed for stars but moving for the Sun, Moon, and planets) can be identified by a declination δ that is the angle above the celestial equator as measured along its hour circle and a right ascension α that is the angle along the celestial equator measured relative to a different hour circle that passes through the vernal equinox.

Figure 1.3 illustrates three great circles on the celestial sphere and the planes that they define. These circles are the celestial equator, the ecliptic (the circle followed by the Sun in its yearly path, and the intersection of the plane of the solar system with the celestial sphere), and the Galactic equator (the intersection of the Galactic plane with the celestial sphere).

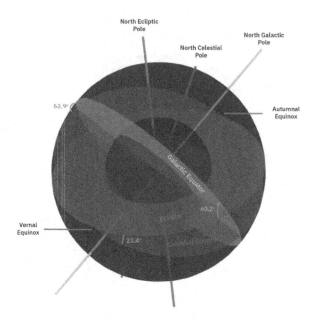

Figure 1.3 An illustration of the three great circles on the celestial sphere, defined by the celestial equator, the ecliptic, and the Galactic equator. Each circle can be used to define a plane in the celestial sphere. The ecliptic can be used to define a north ecliptic pole at the extension of its axis, and it is tilted by 23.4° relative to the north celestial pole. The ecliptic intersects the celestial equator at the vernal equinox (where the Sun is located around March 21) and the autumnal equinox (where the Sun is located around September 21). The relative angles between each pair of circles are also shown. Figure from wikimedia commons, marked as public domain.

Given the inclination of the Galactic plane, the view of the disk of our Galaxy is a bright band of starlight that sweeps across the sky in an arc, usually rising far above the local horizon, as seen in Figure 1.4. This bright band of stars, along with dark lanes of obscuration due to interstellar dust grains, has fascinated mankind since the dawn of humanity. It has been given many names by the various civilizations across the Earth. In Sanskrit, it is *Mandakini*, meaning "unhurried" or "she who flows calmly" and is also the name of a river. A more direct river analogy is used in East Asia, where the Chinese, Japanese, and Korean words translate as "silver river." In Arabic it is *Darb Al-Tabbana*, which translates to "hay merchants way," and the association with straw is also made in many other cultures. In Greek, it translates to "milky circle" and in Latin, it became *Via Lactea* or "milky way." The beauty of the night sky inspires us to explore its wonders and combine astronomical observations with our knowledge of physics gained on Earth to unlock the secrets of the universe.

Figure 1.4 A long-time exposure of the bright band of the Galactic plane as seen from Earth. Credit: Arpan Das, www.arpandas.com.

1.2 UNITS

In physical theories, we are free to choose any independent measures (units) for each of length, time, and mass. All other units can be related to these three units. In astrophysics, the preferred unit system is centimeter-gram-second, or cgs, and we adopt it in this textbook. The cgs system has its advantages over the more commonly used Système International (SI) units that is based on meter-kilogram-second, since it does not introduce additional units for electricity and magnetism. In cgs, the units of electric charge, electric current, electric field, magnetic field, etc. are direct combinations of one centimeter, one gram, and one second. In the SI system, a new unit of Ampere is introduced for the electric current, leading to the unit of Coulomb for electric charge. While these units ease the measurement of laboratory currents, yielding order unity values, we have no such need for it in astrophysical situations, and the cgs system becomes the simpler alternative. Table 1.1 lists the basic cgs units.

For energy levels of atoms and molecules, we often also use the unit of electron volt, or eV, defined by $1\,\text{eV} = 1.60 \times 10^{-12}$ erg. For masses of atoms, we often use the atomic mass unit (amu) defined as one-twelfth of the mass of an atom of ^{12}C, or 1.66×10^{-24} g.

Table 1.2 lists some fundamental physical constants in cgs units. We note that the value of Boltzmann's constant k also defines the unit of Kelvin (K) for temperature. One Kelvin is equal to a change in the thermodynamic temperature T that changes the thermal energy kT by 1.38×10^{-16} erg. The zero

TABLE 1.1 The cgs units

Quantity	cgs unit	expanded
length	cm	cm
mass	g	g
time	s	s
frequency	Hz	s^{-1}
force	dyne	g cm s^{-2}
energy	erg	$\text{g cm}^2\,\text{s}^{-2}$
electric charge	esu	$\text{g}^{1/2}\,\text{cm}^{3/2}\,\text{s}^{-1}$
current	esu s^{-1}	$\text{g}^{1/2}\,\text{cm}^{3/2}\,\text{s}^{-2}$
current density	$\text{esu s}^{-1}\,\text{cm}^{-2}$	$\text{g}^{1/2}\,\text{cm}^{-1/2}\,\text{s}^{-2}$
electric potential	statvolt	$\text{g}^{1/2}\,\text{cm}^{1/2}\,\text{s}^{-1}$
electric field	statvolt cm^{-1}	$\text{g}^{1/2}\,\text{cm}^{-1/2}\,\text{s}^{-1}$
magnetic field	Gauss	$\text{g}^{1/2}\,\text{cm}^{-1/2}\,\text{s}^{-1}$

TABLE 1.2 Fundamental constants

Name	Symbol	Value
Speed of light	c	$3.00 \times 10^{10}\ \text{cm s}^{-1}$
Gravitational constant	G	$6.67 \times 10^{-8}\ \text{g}^{-1}\,\text{cm}^3\,\text{s}^{-2}$
Planck's constant	h	$6.63 \times 10^{-27}\ \text{erg s}$
Electron charge	e	$4.80 \times 10^{-10}\ \text{esu}$
Electron mass	m_e	$9.11 \times 10^{-28}\ \text{g}$
Proton mass	m_p	$1.67 \times 10^{-24}\ \text{g}$
Boltzmann's constant	k	$1.38 \times 10^{-16}\ \text{erg K}^{-1}$
Stefan-Boltzmann constant	σ	$5.67 \times 10^{-5}\ \text{erg cm}^{-2}\,\text{s}^{-1}\,\text{K}^{-4}$
Radiation constant	a	$7.56 \times 10^{-15}\ \text{erg cm}^{-3}\,\text{K}^{-4}$

point of the Kelvin scale is equal to -273 on the Celsius scale. At times, we may also use the constant $\hbar \equiv h/(2\pi)$.

One of the most beautiful features of astrophysics is that it allows one to put together the four major streams of physics: electromagnetism, gravity, quantum mechanics, and thermodynamics, represented by the fundamental constants c, G, h, and k, respectively. In which other field of physics is that possible? We will see in later chapters that the calculation of critical masses like the Jeans mass and the Chandrasekhar mass represents intriguing combinations of some of these constants. Table 1.3 lists some additional constants that are commonly used in astronomy.

Astronomy is the study of both the very small (e.g., atoms, molecules, wavelengths of radiation) and the very large (e.g., stellar, galactic, and cosmic scales); hence, we frequently employ prefixes such as the ones listed in Table 1.4. The general convention is to put the first letter of each prefix in front

TABLE 1.3 Astronomical constants

Name	Symbol	Value
Astronomical Unit	AU	1.50×10^{13} cm
Parsec	pc	3.09×10^{18} cm
Year	yr	3.16×10^{7} s
Solar mass	$M_\odot$	1.99×10^{33} g
Solar radius	$R_\odot$	6.96×10^{10} cm
Solar luminosity	$L_\odot$	3.83×10^{33} erg s^{-1}
Jupiter mass	M_{Jup}	1.90×10^{30} g
Earth mass	$M_\oplus$	5.98×10^{27} g
Earth radius	$R_\oplus$	6.37×10^{8} cm

TABLE 1.4 Common prefixes

Prefix	Factor	Prefix	Factor
pico	10^{-12}	kilo	10^{3}
nano	10^{-9}	mega	10^{6}
micro	10^{-6}	giga	10^{9}
milli	10^{-3}	tera	10^{12}
centi	10^{-2}	peta	10^{15}
deci	10^{-1}	exa	10^{18}

of a unit and sometimes abbreviate the unit as well, e.g., nanometer is nm. For micrometer, however, we use a Greek symbol to get μm. For prefixes of mega and larger, we use a capital letter, e.g., megayear is Myr and petahertz is PHz.

1.3 BASIC PHYSICS FOR ASTROPHYSICS

Here, we review several areas of physics that are essential in astrophysics. A reader may prefer to skip ahead to the next Chapters and use this section as a reference for physical concepts and equations that are presented there.

1.3.1 The Electromagnetic Spectrum

Central to astrophysics is the concept of electromagnetic waves, which provide us with nearly all of our information about the heavens (neutrinos and gravitational waves are the other means). In 1865, James Clerk Maxwell (1831 - 1879) realized that time-varying electric and magnetic fields produced electromagnetic waves that traveled at a speed c. Albert Einstein (1879 - 1955) later realized, through the theory of special relativity, that the speed c is the same for all observers, regardless of their motion relative to the source of the waves.

He later realized that electromagnetic waves also had a complementary avatar as a collection of discrete packets, or quanta, each with energy $E = h\nu$. This seemingly contradictory definition assigns a discrete energy E to a **photon** that is also characterized by the classical concept of frequency ν of an infinite wave train. What sense can one make of all this? That mysterious contradiction led to the development of quantum mechanics. In this book, we only touch upon that story in order to utilize results as needed.

What we regard as the visible spectrum is really just a narrow range of wavelengths of electromagnetic radiation that can penetrate the Earth's atmosphere and reach its surface. Some portion of the infrared, as well as large portions of the radio spectrum, can also reach the Earth's surface. Table 1.5 shows the different parts of the electromagnetic spectrum, listed in terms of wavelength λ, frequency $\nu = c/\lambda$, and photon energy $E = h\nu$.

TABLE 1.5 The electromagnetic spectrum

Name	Wavelength	Frequency	Photon energy
Gamma ray	< 0.02 nm	> 15 EHz	> 62.1 keV
X-ray	0.01 nm − 10 nm	30 PHz − 30 EHz	124 eV − 124 keV
Ultraviolet	10 nm − 400 nm	750 THz − 30 PHz	3 eV − 124 eV
Visible light	400 nm − 750 nm	400 THz − 750 THz	1.7 eV − 3 eV
Infrared	750 nm − 1 mm	300 GHz − 400 THz	1.24 meV − 1.7 eV
Microwave	1 mm − 1 m	300 MHz − 300 GHz	1.24 μeV − 1.24 meV
Radio	1 m − 100 km	3 kHz − 300 MHz	12.4 feV − 1.24 μeV

1.3.2 Specific Intensity and Optical Depth

An important concept in the theory of radiative transfer is the **monochromatic specific intensity** I_ν. This is the electromagnetic power (energy per unit time), with frequencies in the range $[\nu, \nu+d\nu]$, propagating in a particular direction within a solid angle $d\Omega$.

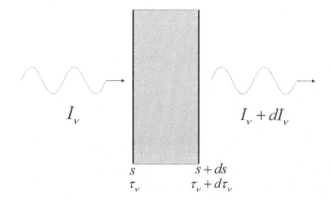

Figure 1.5 An illustration of the geometry of radiative transfer.

Consider a beam of radiation with intensity I_ν that is entering a slab of material, as illustrated in Figure 1.5. If s measures the path length along the direction of propagation and we ignore processes that can change the direction of some of the radiation, then the evolution of the intensity can be described by the equation of radiative transfer:

$$dI_\nu = -I_\nu \, k_\nu \, ds + j_\nu \, ds. \tag{1.1}$$

Here, k_ν is the **attenuation coefficient** at frequency ν and has units of inverse length. It is in fact equal to the inverse of the mean free path, $\ell = 1/(n\sigma)$, where n is the number density of absorbers and σ is the cross section for absorption. A quantity related to k_ν that is used by astronomers is the **opacity**, $\kappa_\nu \equiv n\sigma/\rho$, where ρ is the mass density, so it is effectively the cross section per unit mass of absorbers. Furthermore, j_ν is the **emissivity** at frequency ν and has units of energy per unit time per unit volume per unit frequency per unit solid angle.

In the absence of any emissivity, the solution to Equation (1.1) for a spatially constant k_ν is

$$I_\nu(s) = I_0 \exp(-k_\nu \, s), \tag{1.2}$$

where I_0 is an initial value of I_ν at $s = 0$. To make further progress, it is convenient to introduce a new independent variable, the **optical depth** τ_ν, defined by

$$d\tau_\nu \equiv k_\nu \, ds. \tag{1.3}$$

Note that it is also possible to define τ_ν with the opposite sign, so that the radiation is propagating in the direction of decreasing τ_ν. We focus here on the convention of using a positive sign in Equation (1.3) and the equation of radiative transfer now becomes

$$dI_\nu = -I_\nu \, d\tau_\nu + S_\nu \, d\tau_\nu, \tag{1.4}$$

where

$$S_\nu \equiv \frac{j_\nu}{k_\nu} \tag{1.5}$$

is called the **source function**. We note here that the general solution to Equation (1.4) is

$$I_\nu(\tau_\nu) = I_\nu(0)\, e^{-\tau_\nu} + \int_0^{\tau_\nu} e^{-(\tau_\nu - \tau_\nu')}\, S_\nu \, d\tau_\nu'. \tag{1.6}$$

The above is a general solution to the equation of radiative transfer, in which S_ν can be a function of position. The solution shows that the intensity I_ν at some optical depth I_ν is the initial intensity $I_\nu(0)$ attenuated by the factor $e^{-\tau_\nu}$ plus another integral over the emission that also contains the effect of attenuation along the path.

1.3.3   Blackbody Spectrum

In 1900, Max Planck (1858 - 1947) achieved one of the great triumphs of theoretical physics by deriving the law for the intensity of radiation emitted at different wavelengths by a very opaque body at some temperature T. We refer to a perfectly opaque body as a **blackbody** even though it would not generally appear black. Planck's statistical-mechanical derivation of the radiation law had to assume that radiation was carried in packets known as photons. This idea laid the groundwork for the development of quantum mechanics, even though Planck himself doubted the physical validity of the quantum concept. The **Planck law** explained longstanding measurements of the spectrum of a blackbody that could not be explained by classical physics.

According to Planck's theory, the specific energy density of radiation (energy per unit volume per unit wavelength between λ and $\lambda + d\lambda$) is

$$u_\lambda d\lambda = \frac{8\pi hc}{\lambda^5} \frac{1}{e^{hc/(\lambda kT)} - 1}\, d\lambda. \tag{1.7}$$

Here, u_λ is compact notation for what is really $u(\lambda, T)$, a function of both λ and T. We can think of u_λ as a probability density function for the energy density of photons as a function of wavelength for a given temperature T. Astronomers find it convenient to work with the related quantity $B_\lambda(T)$ that measures the flux of radiant energy, i.e., energy per unit area per unit time per unit wavelength that is radiated into a unit of solid angle, as illustrated in Figure 1.6. Since the radiation field is isotropic (equal in all directions), the two quantities are related by

$$B_\lambda(T) = \frac{c}{4\pi} u_\lambda(T). \tag{1.8}$$

Hence, we can write

$$B_\lambda(T) = \frac{2hc^2}{\lambda^5} \frac{1}{e^{hc/(\lambda kT)} - 1}. \tag{1.9}$$

This function is the monochromatic specific intensity for a radiation field that is in thermal equilibrium at a temperature T. It can also be written in terms of frequency, $B_\nu(T)$, the energy per unit area per unit time per unit frequency that is radiated into a unit of solid angle, via $B_\lambda(T)d\lambda = B_\nu(T)d\nu$ and $c = \nu\lambda$, yielding

$$B_\nu(T) = \frac{2h\nu^3}{c^2} \frac{1}{e^{h\nu/(kT)} - 1}. \tag{1.10}$$

We note that for a system in **thermodynamic equilibrium**, the specific intensity I_ν introduced in the previous subsection is equal to $B_\nu(T)$. Furthermore, if we assume **local thermodynamic equilibrium** (LTE) in a medium of varying temperature T, we are assuming that each region has equilibrated to a Planck distribution of photon energies at its local temperature T.

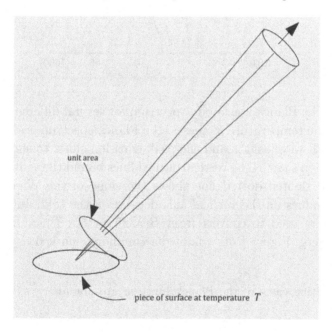

unit area

piece of surface at temperature T

Figure 1.6 The monochromatic specific intensity $B_\lambda(T)$ is the energy per unit time at wavelength λ radiated per unit wavelength per unit solid angle into a small cone from a piece of a blackbody radiator of temperature T. The piece of the blackbody has a unit area in projection that is perpendicular to the axis of the cone. The cone is at a particular angle θ relative to the surface normal and extends a solid angle $d\Omega$.

The Planck function is peaked (see Figure 1.7) with a local maximum at a wavelength $\lambda_{\max}$ that occurs when

$$\lambda_{\max} T = 0.290 \, \text{cm K}. \tag{1.11}$$

Equation (1.11) is known as **Wien's law** and expresses the physical idea that the most probable wavelength of photons in a radiation field in thermal equilibrium at temperature T is found where the photon energy hc/λ is approximately equal to the classical energy per particle $(3/2)kT$.

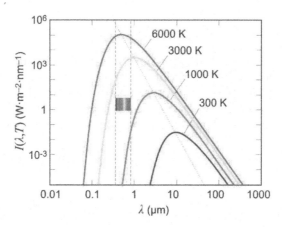

Figure 1.7 The Planck blackbody spectrum for several different temperatures. As the temperature increases, the Planck spectrum radiates more energy at all wavelengths and the peak emission shifts to shorter wavelengths (Wien's law). The vertical dashed lines bound the visible spectral range. The slanted dotted line shows the slope of the Wien distribution. The values on the vertical axis depend on the solid angle $\Delta\Omega$ (at the detector) used to convert from $B(\lambda, T)$ to $I(\lambda, T) = B(\lambda, T)\Delta\Omega$. $1\,\mathrm{W} = 10^7\,\mathrm{erg}$. Figure from wikimedia commons, marked as public domain.

The limiting cases of the Planck function are, for $h\nu \gg kT$, the Wien distribution

$$B_{\nu,\lambda}(T) = \frac{2h\nu^3}{c^2}e^{-h\nu/(kT)} = \frac{2hc^2}{\lambda^5}e^{-hc/(\lambda kT)}, \tag{1.12}$$

and in the low frequency limit, $h\nu \ll kT$, the Rayleigh–Jeans distribution

$$B_{\nu,\lambda}(T) = \frac{2\nu^2 kT}{c^2} = \frac{2ckT}{\lambda^4}. \tag{1.13}$$

The latter is commonly applied in radio astronomy.

Integrating the Planck function yields the total radiant flux (energy per unit area per unit time) leaving a blackbody:

$$F = \sigma T^4. \tag{1.14}$$

This equation is known as the **Stefan–Boltzmann law**.

1.3.4 Level Populations

Consider a collection of atoms that have various internal energy states, and have reached a state of statistical equilibrium at a given temperature T. Here, to avoid confusion with the energy levels of hydrogen denoted by n, we will use N to denote number densities of species. The **Boltzmann equation**, developed by Ludwig Boltzmann (1844 - 1906), relates the level populations N_A and N_B in two energy levels A and B via

$$N_B/N_A = (g_B/g_A) \exp[-(E_B - E_A)/kT], \qquad (1.15)$$

where $g_{A,B}$ represents the multiplicity of an energy level, and $E_{A,B}$ is its energy.

A further insight that has allowed us to ultimately determine abundances of elements in the stars and the universe was made by Meghnad Saha (1893 – 1956). At high temperatures, some atoms become ionized, so the level population relation must account for ionizations as well as recombinations of electrons with ions. A relation describing **ionization equilibrium** then depends on the number densities of atoms, ions, and free electrons. The result is the **Saha equation**,

$$N_+/N_0 = [A(kT)^{3/2}/N_e] \exp(-X_0/kT), \qquad (1.16)$$

where N_+ is the number density of ions, N_0 is the number density of neutral atoms, N_e is the number density of electrons, X_0 is the ionization potential of the atom from the ground state, and A is a constant that incorporates fundamental constants like the electron mass m_e and h, as well as a partition function Z that measures how many ways an atom or ion can arrange its electrons. The value of Z is typically of order unity, and we simplify our treatment here by using the composite constant A.

A practical problem in stellar astronomy is to calculate the fraction of all atoms/ions of a given species that are in a given state. Most famously, the interpretation of stellar spectra requires a calculation of the fraction of all hydrogen that is in the $n = 2$ state. For this purpose, we can combine the Boltzmann and Saha equations. If N is the total number density of hydrogen that is in either a neutral or ionized state and N_2 is the number density in the $n = 2$ state, then the fraction of all atoms/ions in the $n = 2$ state is

$$\frac{N_2}{N} = \frac{N_2}{N_0 + N_+} = \frac{N_2/N_0}{1 + N_+/N_0}. \qquad (1.17)$$

Now in most cases, $N_0 \approx N_1$ since most neutral states are in the ground state. This approximation allows us to simplify the relation to

$$\frac{N_2}{N} = \frac{N_2/N_1}{1 + N_+/N_0}. \qquad (1.18)$$

Equations (1.15) and (1.16) can then be directly inserted into the above equation. Figure 1.8 shows that even though both N_2/N_1 and N_+/N_0 are monotonically increasing functions of temperature, the composite equation (1.18) is peaked with a maximum at $T \simeq 10,000$ K. Physically, this means that when the gas is at temperatures well below 10^4 K, it is mostly neutral and relatively few atoms have been pushed up to the $n = 2$ level through collisions. Furthermore, at temperatures well above 10^4 K, the gas is largely ionized, and therefore, few neutral hydrogen atoms exist, whether in the $n = 2$ state or not.

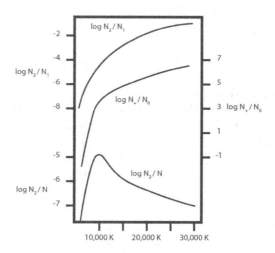

Figure 1.8 The fraction of hydrogen atoms in the $n = 2$ state. The Boltzmann equation yields the relative fraction of atoms in the $n = 2$ state to that in the $n = 1$ state. The Saha equation yields the fraction of ionized atoms. A combination of the two equations yields the fraction in the $n = 2$ state relative to all (neutral or ionized) hydrogen, and reveals a peak at $T \simeq 10,000$ K.

1.3.5 Measuring Temperature

We often think of the concept of temperature in terms of the kinetic motions of microscopic particles in a system. The classical theory of **Maxwell–Boltzmann statistics** predicts that a collection of particles that interact via perfectly elastic short-range collisions will reach an equilibrium distribution function of speeds v that is

$$f(v) = \left(\frac{m}{2\pi kT}\right)^{3/2} 4\pi v^2 \exp[-(1/2)mv^2/(kT)], \tag{1.19}$$

where T is the temperature and m is the mass of each particle. Here, $f(v)$ represents that fractional number of particles that have a speed between v and $v + dv$.

The Maxwell–Boltzmann distribution has root mean square value

$$v_{\rm rms} \equiv \langle v^2 \rangle^{1/2} = \left(\frac{3kT}{m} \right)^{1/2}. \tag{1.20}$$

The above relation provides an intuitive way to understand the concept of temperature T, however, that is only one possible measure of temperature, known as **kinetic temperature**. Below we list the various different measurements of temperature.

Kinetic temperature: determined from the Maxwell–Boltzmann distribution, e.g., from $v_{\rm rms}$ that is often measured by the width of a spectral line.

Excitation temperature: determined using the Boltzmann equation.

Ionization temperature: determined using the Saha equation.

Color temperature: from the shape of the Planck function or Wien's law.

Effective (radiation) temperature: from the Stefan-Boltzmann law.

All these different measures show that there is no unique measure of temperature. However, all measurements should yield the same value if a system is in a state of thermodynamic equilibrium. In other situations, the values can be very different. For example, a low-density gas bathed in a strong radiation field may have a high excitation temperature but not have relaxed to a state where collisions among atoms establish the same kinetic temperature.

1.3.6   Spectral Line Formation

Spectral lines can sometimes appear as an enhancement of emission above the background (continuum) level, in which case they are called **emission lines**, or they can appear as a decrease in the continuum level, in which case they are called **absorption lines**. When can we expect to see one or the other, or see just a continuous spectrum? **Kirchoff's laws** of radiation for gases (or Kirchoff's rules), developed by Gustav Kirchoff (1824 - 1887), help us to make sense of what to expect. Succinctly, these principles are:

(1) A hot and opaque solid, liquid, or highly compressed gas emits a continuous spectrum.

(2) A hot, transparent gas produces a spectrum of emission lines. The specific lines depend on which elements are present in the gas.

(3) Relatively cool, transparent gas in front of a continuum source produces absorption lines. The specific lines depend on which elements are present in the gas.

We can use these principles to understand the spectrum of the Sun and other stars. The interior is extremely opaque and emits blackbody radiation characterized by a surface **effective temperature** T_e, which can differ from star to star. However, the photosphere is actually not infinitely thin, and we can interpret the spectrum as being shaped by cooler atoms in the upper photosphere that absorb out specific lines from the continuum emission emerging from below. So, a typical stellar spectrum is a blackbody characterized by some temperature T_e but with a series of absorption lines corresponding to the atoms or molecules that exist at the star's surface.

We may then ask the question: what happens after a photon is absorbed? Can an absorption line really be formed if the electron quickly de-excites and releases another photon of exactly the same wavelength? The answer is of course yes. We see absorption lines! They are characterized by less intensity that the continuum but are not completely dark so that many photons of that wavelength do reach the observer. The explanation for absorption lines is three fold, as listed below:

(1) De-excitation by collision, so no photon is produced.

(2) De-excitation to a different level, so a photon of a different energy (therefore wavelength) is produced.

(3) De-excitation to the original level, but the photon (of the same energy as the original photon) may be emitted in any direction, so is unlikely to be emitted in the same direction as the original photon.

Emission lines are produced by collisional excitation to a higher level, followed by radiative decay so that a photon of energy $\Delta E = h\nu$ is produced. In astronomy, we sometimes measure **forbidden emission lines**, which are emission lines that would not normally be observed in a terrestrial laboratory, since the transition would be collisionally de-excited. However, in the rarefied regions of space, the de-excitation can occur radiatively and an emission line is created. The most famous forbidden emission line in astronomy is the 21 cm emission line of hydrogen, emitted from rarefied interstellar atomic hydrogen gas clouds.

1.3.7 Line Broadening

Spectral lines originate from the quantum mechanical rules of energy transitions between different energy states of electrons in an atom or molecule. Although quantum mechanics prescribes specific discrete values for the allowable energy gaps, there is always a spread of wavelengths/frequencies at which photons are absorbed, emitted, or detected. This leads to the concept of spectral line width.

The strength of a spectral line can be quantified by the concept of **equivalent width**, which measures the area contained in a spectral line. The concept is illustrated in Figure 1.9. On a plot of intensity versus wavelength, one can draw a rectangle that has a height equal to that of the continuum intensity. If we choose the width of this rectangle such that the area in the rectangle equals the area in the spectral line that is contained above or below the continuum level, then the width of the rectangle is the equivalent width. Ultimately, the equivalent width of a spectral line depends on the number of atoms or ions that are in the energy state from which the transition occurs, but the appearance of the line can depend on many broadening mechanisms.

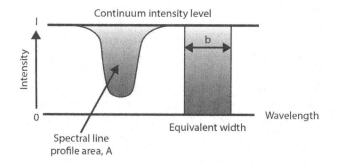

Figure 1.9 An illustration of the concept of equivalent width.

What causes spectral line broadening? A variety of mechanisms are listed below.

Natural Broadening: Quantum mechanics imposes an uncertainty principle that relates energy and time. The electron in a particular energy level has an energy uncertainty ΔE that is related to its lifetime Δt in that level via

$$\Delta E \, \Delta t \gtrsim h.$$ (1.21)

Here, h is used as an order of magnitude value and depending on how the uncertainty or spread of values is defined, some sources quote a value of $\hbar$ or $\hbar/2$ on the right hand side. If the lifetime of an excited level is typically $\Delta t \simeq 10^{-8}$ s, then the quantum relation for the energy of a photon, $E = h\nu$, leads to a frequency uncertainty $\Delta\nu = \Delta E/h = 1/\Delta t$. Using the relation $c = \lambda\nu$ for electromagnetic waves, we find a wavelength spread in the spectral line of

$$\Delta\lambda = \frac{\lambda^2}{c}\Delta\nu.$$ (1.22)

Inserting a representative value $\lambda = 500$ nm, we find that $\Delta\lambda = 10^{-4}$ nm, a truly small effect, which acts as a lower limit to the line broadening.

Collisional (Pressure) Broadening: energy levels can be shifted by the electrostatic interaction of atoms with neighboring atoms. This effect has a direct dependence on particle density. In a gas of sufficient density, characteristic spectral features will disappear entirely, and the gas emits a continuum of wavelengths.

Thermal Doppler Broadening: microscopic atomic motions along the line-of-sight lead to Doppler shifts

$$\frac{\Delta\lambda}{\lambda} = \frac{v}{c}.$$ (1.23)

Here, v is the root mean squared speed of the particles and $\Delta\lambda$ is the dispersion of the spectral line.

Magnetic Fields (Zeeman Effect): an ambient magnetic field causes a splitting of some energy levels into several components. If the splitting is resolved, i.e., several lines are measured instead of one, the amount of splitting can be used to measure the magnetic field strength. However, if the splitting remains unresolved, then one sees a broadened version of the original single spectral line.

Macroscopic Broadening: Doppler shifts from large-scale unresolved motions within the observed region can also cause line broadening. The motions could be due to expansion, contraction, rotation, or turbulence.

1.4   CONCLUDING THOUGHTS

We have reviewed many properties of electromagnetic radiation in this Chapter since it is the means by which we have learned the most about the universe. A grounding in the basic ideas of continuum (blackbody) and spectral line radiation is required to make sense of the vast amounts of data being generated from ground- and space-based observatories in all parts of the electromagnetic spectrum. Quantum mechanics is inherent in astronomy, because the ideas of quanta of radiation and discrete energy levels underpin our understanding about blackbody radiation and spectral line formation. The Saha equation has proven to be essential to unlocking the secret of the actual chemical composition of the Sun. This distribution of elements, dominated by hydrogen and helium, is essentially the same for all stars, galaxies, and intervening visible matter, yielding evidence for an early hot phase of the universe, the **Big Bang**. In recent times, neutrino astronomy and gravitational wave astronomy have added to the repertoire, enabling studies of the Sun's interior, supernovae, and black hole mergers. Furthermore, the *absence* of electromagnetic radiation has been used to infer the presence of dark matter, where the acceleration of visible matter is explained by the gravitational effects of the dark matter. Some of the best evidence for dark matter however comes from radiation itself, in the **lensing** (bending and focusing of light) by galaxy clusters of the light from background objects. Astronomy offers a rich and never-ending exploration of the consequences of basic physics!

CHAPTER 2

The Interstellar Medium

2.1 INTRODUCTION

The interstellar medium (ISM) is composed of all the mass–energy between the stars, including gas, dust, cosmic rays, magnetic fields, and radiation. It accounts for some of the most beautiful images in astronomy, including colorful interstellar emission and reflection nebulae, as well as dark filaments of dense and dusty gas within which new stars are born. The plasma state of partially or fully ionized gas is the most common form of ordinary matter in the universe. This Chapter focuses on the ISM, but many of the principles described here can also be used in the study of stellar interiors and the hot gas that comprises the intergalactic medium (IGM). Interestingly, most of the visible matter in the universe is thought to be in the form of hot diffuse IGM gas.

DOI: 10.1201/9781003215943-2

Figure 2.1 A multiwavelength composite image of the diffuse nebula NGC 6357. This object is also known as the Lobster Nebula due to its shape when viewed in visible light on larger scales than seen here. The nebula contains young stars and expanding bubbles of hot gas surrounding the massive ones. X-ray emission from hot gas is colored purple, infrared emission from dust is colored orange, and the optical light includes blue light from massive stars that is reflected by dust. Credit: X-ray from NASA/CXC/PSU/Townsley et al. (2014); Optical from UKIRT; Infrared from NASA/JPL-Caltech.

It is hard to believe that there was no awareness of the ISM till the early 20th century. Dark patches in the sky, or dark nebulae, were known in the 19th century. However, it was a matter of debate whether they represented local patches of obscuring material (a dark cloak) or whether they were holes in the sky where no stars existed. The holes would of course have to be more like deep tunnels aligned with our line of sight in order to not reveal any background stars. Evidence for an ISM emerged in the early 1900s, when Johannes Hartmann (1865–1936) observed that there were stationary narrow absorption lines in the spectra of binary star systems, implying the existence of an intervening medium between the system and the Earth. Further work by Plaskett and Pierce in the 1930s placed this interpretation on a firmer footing by demonstrating that the strength of these absorption lines correlated with distance, with stronger lines from more distant stars due to more intervening ISM. In 1930, Robert Trumpler (1886–1956) demonstrated the existence of interstellar dust by measuring the apparent diameter and brightness of star clusters. The dimming effect could be explained by an intervening absorbing medium that extinguished starlight. In the 1940s and 1950s, the advent of radio astronomy led to the prediction and detection of widespread 21 cm emission from atomic hydrogen gas in the ISM. Molecular lines from OH, CO, and many other species were detected from the 1960s, establishing the existence of clouds of molecular gas. Large-scale magnetic fields were detected through radio observations of the **Zeeman effect** (splitting or broadening of spectral lines due to an ambient magnetic field), and indirectly through polarization of starlight, **synchrotron radiation** (radiation from a relativistic charged particle as it is accelerated in a spiral path about a magnetic field), and **Faraday rotation** (the systematic rotation of the polarization direction of radiation as it passes through a plasma). A population of high energy particles (cosmic rays) were also identified through the synchrotron radiation.

Altogether, the ISM is a beautiful, complex, and highly important component of galaxies that consists of gas, dust, magnetic field, cosmic rays, and the interstellar radiation field. The ISM is also a stellar nursery that provides the fundamental conditions for star birth in some regions. When considering the evolution of the universe as a whole, we find that the formation and evolution of galaxies is also a story of the transformation of their ISM into stars. The efficiency of this transformation at different epochs determines in large part the history of the universe.

2.2 THE DISCOVERY OF INTERSTELLAR DUST

Mixed in with all but the hottest gas is interstellar dust. Dust grains are an important source of interstellar extinction, gas-phase element depletion, sites of interstellar chemistry, etc. They are a solid phase rather than gas-phase material.

Dust grains range in size from a few microns down to macromolecular scales (clumps of 50–100 atoms or less). The role of dust is indeed disproportionate to its share of the mass of the ISM. They are the primary cause of visual extinction in astronomical observations, and play a crucial role in the chemistry of the interstellar medium, the formation of molecules, the heating of the ISM, and in setting the ionization fraction in interstellar gas. The latter is crucial to determining the coupling of gas to the interstellar magnetic field.

Dust was first inferred through its effects of **extinction** and **reddening** of starlight. Extinction is the dimming of light as it makes its way toward the observer. Reddening is a measure of how much the dimming is a function of wavelength, with shorter wavelengths of light usually extinguished preferentially. An everyday example is the redness of the Sun at sunrise or sunset, when the sunlight travels through the maximum length of atmosphere before reaching our eyes. Extinction itself can occur due to either direct **absorption** or **scattering** of light. Here, we define absorption as the process by which at least some of the energy of an incoming photon is extracted by matter (atoms, molecules, or dust grains), and scattering as a process by which the direction of an incoming photon is changed, without the extraction of energy.

Trumpler inferred the existence of dust by showing evidence for absorption of light from distant open clusters (Trumpler, 1930). Open clusters are found in the Galactic plane and have similar sizes and numbers of stars. Assuming that they are drawn from distributions with mean diameter D and luminosity L, he reasoned that the distances to the clusters r as determined from the measured apparent angular size $\theta = D/r$ and from the flux $f = L/(4\pi r^2)$ should be correlated with one another. One might expect a straight-line correlation on a log–log plot, with scatter due to variations in D and L about mean values. Figure 2.2 shows the result that there is instead a systematic deviation, implying an intervening obscuring material whose effect increases as the distance r to the object increases.

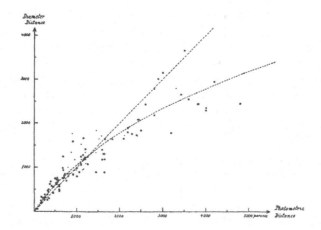

Figure 2.2 Evidence for extinction. The correlation of the distance determined from the angular size with the distance determined from the radiant flux breaks down for distant objects. Credit: Trumpler (1930), reproduced with permission from the Astronomical Society of the Pacific.

Trumpler also detected reddening due to scattering of short wavelengths, similar to what happens in the Earth's atmosphere. Figure 2.3 illustrates the mechanisms of extinction and reddening of the observed starlight.

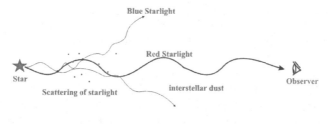

Figure 2.3 Illustration of interstellar reddening due to scattering by dust.

Although its effect through extinction is very significant, obscuring our view toward the Galactic center and creating localized patches of darkness (dark clouds), the mass of dust is estimated to be only about one percent of that of the interstellar gas. The interstellar dust is thought to originate from condensation within winds from cool supergiant stars, and is likely comprised of some combination of graphite and silicates (some of which have icy mantles of sizes $\sim 0.005 - 0.2$ mm). Figure 2.4 illustrates the evidence for this composition. There are also the polycyclic aromatic hydrocarbons (PAHs), which are very long chain molecules with at least 50 atoms, mostly carbon arranged in stable rings, but with hydrogen atoms along the periphery. They share many characteristics with dust particles.

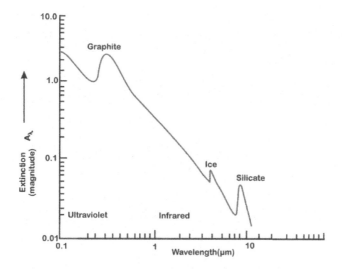

Figure 2.4 Schematic interstellar extinction curve, illustrating the graphite peak at around 0.2 μm and infrared spectrum illustrating the prominent absorption bands due to ices and silicates.

2.3 THE DISCOVERY OF INTERSTELLAR GAS

The discovery by Hartmann (1904) of narrow stationary absorption lines in the spectrum of some binary star systems laid the foundation for the discovery of interstellar gas. The lines were stationary with respect to the cyclical orbital Doppler shifts of the stellar lines seen in the binary star system. It was later found by Plaskett & Pearce (1930) that the strengths of these absorption lines correlated with distance. Lines from more distant stars were stronger. This implied the existence of an intervening interstellar gas. In 1944, Henrik van de Hulst (1918–2000) predicted that the 21 cm emission line of H would be widespread in the ISM, and it was subsequently detected by Ewen & Purcell (1951) and others. This line originates from the magnetic interaction between the electron and proton spins. When the electron is in the $n = 1$ ground state of hydrogen, there is a slightly greater energy when the spins are parallel than when they are antiparallel. The small energy gap ΔE between the two corresponds to a wavelength $\lambda = 21$ cm using the quantum relation $\Delta E = hc/\lambda$. Since ΔE is small, the electrons can be pushed up to the higher level through collisions even in the very cold depths of interstellar space. Although the excitations and de-excitations are rare events, the electron transition back to the lower hyperfine level results in widespread 21 cm emission throughout the universe, wherever there is neutral hydrogen gas. The process is illustrated in Figure 2.5.

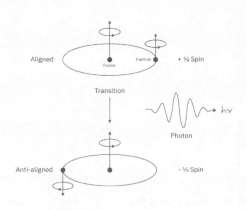

Figure 2.5 Illustration of the hyperfine spin-flip transition of an electron in the $n = 1$ ground state that leads to the 21 cm line of hydrogen. The antiparallel spins of the electron and proton constitute the lower energy state.

Figure 2.6 illustrates how the sky lights up when viewed in 21 cm, even with a few seconds of observation on a modern radio telescope. The dark patches that are seen in optical observations, due to extinction by dust near the Galactic plane, are replaced by bright emission from interstellar gas at radio or submillimeter wavelengths.

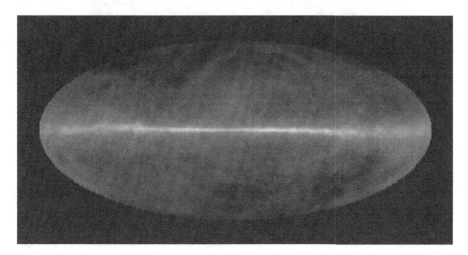

Figure 2.6 An all sky map of 21 cm emission from data in Dickey & Lockman (1990), highlighting the concentration of gas near the Galactic plane and also some arcs of emission that are attributable to stirring by stellar activity. Credit: J. Dickey, F. Lockman.

It turns out that gas of sufficient density allows the formation of molecules, H_2, and trace molecules like CO, CN, OH, etc. that were detected by their spectral lines since the mid-1960s. The cool molecular gas and the dust are distributed in an even thinner layer than the atomic hydrogen gas that is observed in 21 cm.

2.4 AN INVENTORY OF THE INTERSTELLAR MEDIUM

2.4.1 Gaseous Nebulae

The astronomer William Herschel (1738–1822) described dark nebulae as *ein loch in Himmel*-"holes in the star clouds." These holes were dark patches in optical observations that are now known to hide more distant objects. They contain gas and dust where dust blocks visual wavelengths. Figure 2.7 shows one such dark cloud, the B68 globule. In such clouds, the cool gas emits in wavelengths longer than the visible range.

Figure 2.7   Optical (left) and near-infrared (right) images of the B68 glob-ule, illustrating full absorption of background starlight at optical wave-lengths and only partial absorption in the near-infrared, so that the background stars become visible. Credit: Alves et al. (2001), reproduced with permission from Springer Nature.

Among gaseous nebulae, there are various subtypes as listed below. Ex-amples of these objects are also shown in Figures 2.8, 2.9, and 2.10:

Reflection Nebulae: in these types of nebulae, light is scattered (mainly by dust) from an embedded star. Blue light, whose wavelength is closest to the typical size ($\sim 0.1\mu m$) of the dust grains, is preferentially scattered.

Emission Nebulae: in these nebulae, the emission is observed from recom-bining atoms (usually H) within zones of ionized material (H II regions). These regions are created by hot (O or B type) stars that emit copious amounts of Lyman continuum photons that can ionize H atoms. We often observe the red $H\alpha$ line emitted during the recombination cascade.

Planetary Nebulae: these are quite similar to emission nebulae, and the object responsible for excitation is a hot evolved star. The material expelled from star in the post-main-sequence phase forms the nebula. These are among the most visually spectacular objects in the sky.

Supernova Remnants: these are seen in optical emission from ionized ma-terial and radio emission from relativistic electrons spiraling around magnetic fields. We observe a supernova remnant when a shock wave sweeps up sur-rounding gas as it travels outward. The supernova remnant is the aftermath of this shock wave and consists of compressed, heated, and ionized interstellar gas. Figure 2.10 shows the famous Crab Nebula, the remnant of a supernova that was visible to the naked eye and recorded historically in 1054 AD by several cultures around the world. What is seen in that part of the sky is a fast-moving gaseous remnant that encloses a pulsar (neutron star). The rem-nant is accelerating outward, so in the non-inertial frame of reference of the swept-up shell, there is an effective gravitational field pointing back inward toward lower density gas. This creates conditions for a **Rayleigh–Taylor in-**

stability, in which fingers of dense gas will fall back down along the direction of an effective gravitational field, and into the low density region.

Figure 2.8 The Trifid Nebula, containing emission nebulae (pink), reflection nebulae (blue), and filamentary dark nebulae. Credit: Adam Block, AdamBlockPhotos.com.

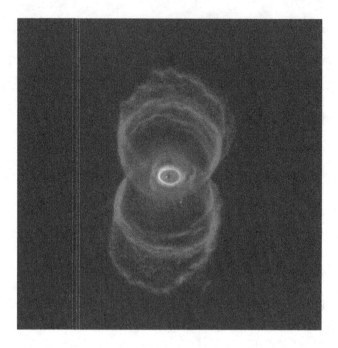

Figure 2.9 The Hourglass Nebula is an example of a planetary nebula. Created by the expelled gases from a dying star, they often have ring-like or bubble-shaped structures but also can be shaped by rotation or the surrounding medium into very complex structures. Credit: R. Sahai and J. Trauger (JPL), the WFPC2 science team, and NASA/ESA.

Figure 2.10 The Crab Nebula supernova remnant. This remnant has a pulsar at the center and the outward moving shell shows the development of the Rayleigh–Taylor instability in the post-shock region. Credit: J. Hester and A. Loll, NASA, ESA.

2.4.2 Interstellar Gas

The advent of radio astronomy from the late 1940s showed that the interstellar gas is spread throughout the Galaxy, and not just confined to the bright nebulae seen in optical observations. We now view the nebulae as simply the brightest (or darkest) spots in a vast collection of gas in the Galaxy. We can conceptually divide the gas into four "phases." The easiest to observe is neutral atomic hydrogen, due to the ubiquity of its 21 cm emission. The neutral hydrogen accounts for two of the four distinct phases, each of which is briefly described below.

Warm Neutral Medium (WNM): This neutral hydrogen (H I) gas has a mean number density $n \simeq 0.2 - 0.5$ cm^{-3} and temperature $T \simeq 8000$ K. It is generally seen in emission of the 21 cm line.

Cold Neutral Medium (CNM): Seen primarily through absorption lines in 21 cm, this phase of neutral hydrogen exhibits narrow line widths that are consistent with a density in the range $20 - 50$ cm^{-3} and temperature $T \simeq 80$ K.

Coronal Gas: This component was discovered in the 1970s by space-based ultraviolet and X-ray telescopes, and consists of hot gas with $T \simeq 10^6$ K at very low density $\simeq 10^{-3}$ cm^{-3}. It is thought to have been shock-heated by supernova remnants and accounts for an unknown but possibly very large-volume fraction of the ISM.

Molecular Clouds: Consisting primarily of H_2 but primarily detected in emission from trace molecules such as CO, OH, and NH_3, molecular clouds have significant self-gravity and maintain a higher internal pressure than the rest of the ISM, with mean densities $10^2 - 10^4$ cm^{-3} and temperatures $T \simeq 10 - 50$ K. Significant line emission and consequent cooling brings the temperature down to these levels.

Table 2.1 summarizes the key properties of the different phases of the ISM. Some descriptions of the ISM also include a **Warm Ionized Medium** that refers to a widespread diffuse ionized gas that is detected primarily by the effect of free electrons that cause the dispersion of radio pulses that come from pulsars. Of course, significant levels of ionization also exist in the coronal gas and emission nebulae.

TABLE 2.1 Phases of the ISM

Phase	Probe	n (cm^{-3})	T (K)
Coronal gas	UV, X-ray	10^{-3}	10^6
WNM	21 cm emission	$0.2 - 0.5$	8000
CNM	21 cm absorption	$20 - 50$	80
Molecular Clouds	CO, OH, NH_3,...	$10^2 - 10^4$	$10 - 50$

An important advance in the theoretical understanding of the ISM was made in the 1960s by Field, Goldsmith, and Habing (Field et al., 1969), who realized that the CNM and WNM acted like two phases (e.g., like liquid water and water vapor at the Earth's surface) that can coexist at the same ambient pressure (usually measured in units of $P/k = nT$) but at different densities and temperatures. The model is illustrated in Figure 2.11. The later detection of widespread hot coronal gas led to the idea of a three-phase medium, each of which exists in pressure balance with the others. The coronal gas is attributed to the massive stars in the Galaxy that create large bubbles of hot gas in the post-shock regions behind supernova remnants. For the atomic gas, the primary heating sources are the photoelectric effect from dust grains, driven by Galactic background ultraviolet starlight, and collisions with cosmic rays. Cooling is primarily due to atomic line cooling, with notable contributions from singly ionized carbon, C II.

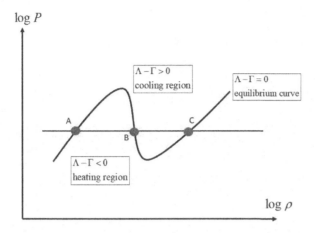

Figure 2.11 An illustration of the two-phase model for atomic hydrogen as determined by Field et al. (1969). The equilibrium curve is shown for which the rates of cooling Λ and heating Γ are balanced. For gas that is placed at a point above the curve, the cooling dominates and the temperature decreases, while the opposite holds for gas at a point below the curve. For interstellar gas at a particular pressure highlighted by the horizontal line, gas can exist in equilibrium at points A, B, and C. However, points A and C represent stable equilibria at which the pressure increases with increasing density along the equilibrium curve, while point B represents an unstable equilibrium and is not expected to be found in the interstellar medium. Points A and C correspond to the two phases of WNM and CNM, respectively.

2.4.3 Magnetic Field

In 1949, Hall and Hiltner independently (Hiltner, 1949; Hall, 1949) discovered that light from most stars is linearly polarized. It was soon accepted that this was due to intervening dust, as the amount of polarization was correlated with the amount of reddening, which was known to be caused by interstellar dust. This was backed up by the theory that there would be alignment of elongated paramagnetic dust grains with their long axis perpendicular to the local magnetic field (Davis & Greenstein, 1951). Later work (Mathewson & Ford, 1970) showed that the inferred magnetic field was largely parallel to the direction of the Galactic plane, but with notable excursions. In 2016, the Planck mission presented a new more comprehensive view of the Galactic magnetic field (see Figure 2.12), this time based on dust emission rather than absorption.

Figure 2.13 illustrates the two means of detecting large-scale magnetic fields in the cosmos through polarized light.

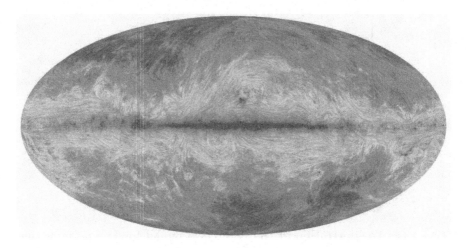

Figure 2.12   An all-sky map of gas column density inferred from 353 GHz dust emission overlaid with the inferred magnetic field directions over-laid as a "drapery" pattern. The magnetic field direction is assumed to be orthogonal to the observed direction of partial polarization of the emission. Credit: ESA and the Planck collaboration.

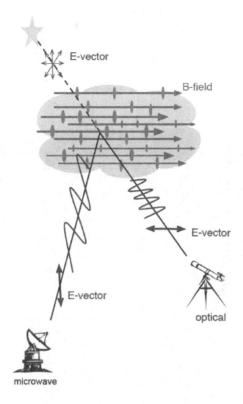

Figure 2.13 Mechanisms of polarized absorption and emission by dust grains that have their long axis perpendicular to the ambient magnetic field direction. Credit: Tassis et al. (2018), reproduced with permission.

So why is there a large-scale magnetic field in our Galaxy, and for that matter in nearly all astrophysical bodies from planets to stars to galaxies? See Figure 2.14. Why is there not a large-scale electric field? Students of electromagnetism are often taught that there is a symmetry between electric and magnetic fields, and that for any inertial reference frame in which there is a magnetic field, one can transform to another inertial frame in which there is an electric field instead. However, this is only part of the story. Maxwell's equations of electromagnetism show that there is an asymmetry between electric and magnetic fields concerning their material sources. Monopole sources of electric field are ubiquitous, but monopole sources of the magnetic field do not exist, or are exceedingly rare! Consider the frame of reference of the center of mass of a body of plasma, e.g., the Sun, a molecular cloud, or the Galaxy. There are free electric charge carriers but no magnetic counterparts. The high conductivity of astrophysical plasmas means that free electric charges, which come in equal quantities of plus and minus charges, can move quickly to short out large-scale electric fields. On the other hand, no free magnetic charges (monopoles) exist that can short out large-scale magnetic fields. Therefore, if electric currents can set up large-scale magnetic fields, they may persist for a long time due to the high conductivity of the plasma. It turns out that only very small relative motions between electrons and ions produce the necessary conduction currents to account for the magnetic field strength of astrophysical plasmas.

Figure 2.14 Magnetic fields in the face-on galaxy NGC 1068 are revealed by polarization of emission from magnetically-aligned dust measured in the far-infrared band at 89 μm using the *Stratospheric Observatory for Infrared Astronomy* (SOFIA). The direction of the inferred magnetic field is illustrated by the drapery pattern, overlaid on a visible light and X-ray composite image of the galaxy from the *Hubble Space Telescope*, *The Nuclear Spectroscopic Array*, and the *Sloan Digital Sky Survey*. The magnetic field shows a global ordering and is generally aligned with the spiral arms of the galaxy. Credit: NASA/SOFIA; NASA/JPL-Caltech/Roma Tre Univ.

One of the most intriguing properties of magnetic fields is inductance, the inducement of electric currents when there is a changing magnetic field. This was discovered in laboratory experiments by Michael Faraday (1791–1867) in 1831. In the low-density interstellar plasma, the electrical conductivity is very high and the induced currents are successful in preserving the magnetic flux

$$\Phi = \int_S \mathbf{B} \cdot d\mathbf{S} \tag{2.1}$$

through a surface S in any Lagrangian displacement of S. This phenomenon is known as "flux freezing," as one may conceptually regard the magnetic field lines as moving with the fluid, like strings embedded in a fluid.

We can carry the string analogy even further and start thinking of field lines as exerting a tension and pressure that resists distortion. So, the strings become more like elastic rubber bands. This arises from the expression for the magnetic Lorentz force per unit volume $\mathbf{f}_L = (\mathbf{j} \times \mathbf{B})/c$, where $\mathbf{j}$ is the electric current density, $\mathbf{B}$ is the magnetic field vector, and c is the speed of light. In the limit of negligible displacement current, which is valid for the long timescale phenomena in the interstellar medium, we can use **Ampere's law** $(\mathbf{j} = (c/4\pi)\nabla \times \mathbf{B})$ and vector identities to find

$$\mathbf{f}_L = -\nabla \frac{\mathbf{B}^2}{8\pi} + \frac{1}{4\pi}(\mathbf{B} \cdot \nabla)\mathbf{B}. \tag{2.2}$$

This can be further simplified to

$$\mathbf{f}_L = -\nabla_n \frac{\mathbf{B}^2}{8\pi} + \hat{n}\frac{\mathbf{B}^2}{4\pi R_c}, \tag{2.3}$$

where $\hat{n}$ is the local normal to the field line and pointing toward the center of curvature, and R_c is the radius of curvature of the field line. See Figure 2.15. The first term in the above equation is a force due to a gradient of magnetic pressure, in analogy to the force due to a pressure gradient $(-\nabla P)$ in a fluid, and the second term is a tension force that is inversely proportional to the local radius of curvature of a field line, in analogy to the force due to tension in a curved string.

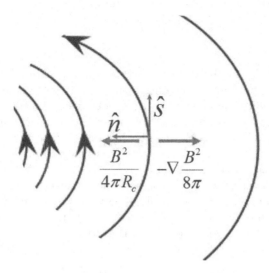

Figure 2.15 Dipolar magnetic field lines. The magnetic pressure gradient force and magnetic tension force can be calculated for such lines in terms of a local coordinate system with normal unit vector $\hat{n}$ and tangential unit vector $\hat{s}$.

The above intuitive ideas of magnetic flux freezing, magnetic pressure gradient, and magnetic tension lead to a picture of magnetic fields where we can think of them loosely as rubber bands carried along with and reacting to the flow of a plasma. An interesting outcome of these effects is the development of an hourglass magnetic field configuration due to the gravitational contraction of gas in a medium with initially near-uniform magnetic field. The central condensed region will have a pinched magnetic field configuration due to magnetic flux freezing, and the concepts of magnetic pressure and tension can be used to see that the magnetic force will resist the gravitational contraction. See Figure 2.16.

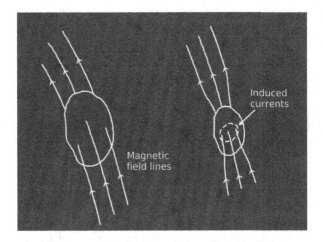

Figure 2.16 The principle of magnetic flux freezing in a highly conducting plasma. If a region of plasma is compressed, then self-induced currents will generate an additional local magnetic field so that the flux through any contour is preserved as it undergoes distortion.

Another interesting consequence of flux freezing is that a plasma with an embedded magnetic field will, when perturbed, develop perturbations in the magnetic field vector as well. Figure 2.17 illustrates the bending of a magnetic field line and the consequent magnetic tension force that acts to bring the plasma and magnetic field back to its unperturbed position.

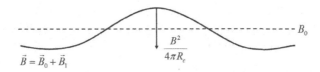

Figure 2.17 The distortion of a large scale magnetic field due to a sinusoidal perturbation in the case of flux freezing. The magnetic tension force is illustrated and acts to bring the plasma and magnetic field back to its unperturbed position, resulting in the propagation of Alfvén waves.

The restoring force due to the magnetic field of strength B and the inertia of the plasma, represented by its mass density ρ, work together to allow a propagating transverse wave along the direction of the background magnetic field. These waves are known as Alfvén waves, named after the noted plasma physicist Hannes Alfvén (1908–1995), and propagate at the Alfvén speed $v_A \equiv B/\sqrt{4\pi\rho}$.

In the 1950s, Davis (1951) and Chandrasekhar & Fermi (1953) realized that the observations of fluctuations in the mean polarization direction of starlight could be equated with the dispersion of magnetic field directions due to propagating Alfvén waves, and be used to estimate the magnetic field strength. The theory of propagation of linear incompressible Alfvén waves leads to the relation

$$\frac{\delta v}{v_A} = \frac{\delta B}{B}, \tag{2.4}$$

where δv is the velocity dispersion of the plasma, v_A is the Alfvén wave speed defined earlier, δB is the fluctuation amplitude of the magnetic field, and B is the background magnetic field strength. The polarization directions reveal a dispersion in angular direction $\delta\theta$ that can be equated to $\delta B/B$, and then Equation (2.4) can be rearranged to yield

$$B = \sqrt{4\pi\rho}\,\frac{\delta v}{\delta\theta}. \tag{2.5}$$

Observations of spectral line emission can yield δv and also be used to estimate ρ, and the polarization observations yield $\delta\theta$. Using observational estimates of these quantities in the Galactic spiral arm, Chandrasekhar & Fermi (1953) estimated a magnetic field strength of $7\,\mu G$. This estimate using the **Davis–Chandraskehar–Fermi method** is very close to the accepted value today, determined using a variety of additional means such as Faraday rotation and the Zeeman effect.

2.4.4 Cosmic Rays

Cosmic rays are highly energetic nuclei and electrons that reach the Earth from all directions. They were originally inferred as the mechanism to explain increasing levels of ionization above the Earth's surface. They were confirmed by Victor Franz Hess (1883 - 1964) in 1912, using a balloon borne electrometer to confirm that the rising ionization required a source that was above the atmosphere. He ruled out the Sun as a possible source by performing the experiment during a near-total solar eclipse. Later work established that the ionizing source was energetic particles and not radiation. The energy spectrum of cosmic rays is very broad, and has a power-law distribution that has values as high as $\sim 10^{21}$ eV.

How then are they confined to the Galaxy? While free particles moving with $v \sim c$ should escape the Galaxy, the cosmic rays instead enter into a symbiotic relationship with Galactic interstellar gas (plasma) and the Galactic gravitational field. The gas is confined by gravity and maintains currents that generate the Galactic magnetic field, whereas cosmic rays are confined by spiraling about the magnetic field lines, but not by gravity. Cosmic rays are

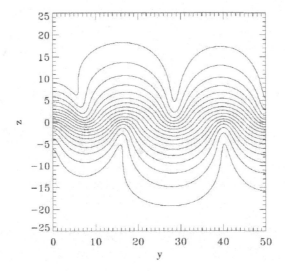

Figure 2.18 Magnetic field line structure along a vertical cut through the Galaxy. Here the y-direction is along a local spiral arm of the Galaxy and parallel to the Galactic plane, which is the horizontal line $z = 0$. From a numerical simulation of the development of the Parker instability. Credit: Basu et al. (1997), reproduced with permission from the American Astronomical Society (AAS).

also a source of ionization of interstellar gas, and thus determine the level at which the plasma can respond to magnetic forces; the greater the ionization fraction, the greater the plasma is coupled to the magnetic field.

The magnetic field lines, while nearly parallel to the Galactic plane at small heights, are expected to have large loop-like excursions above and below the Galactic plane (see Figure 2.18). This is partially due to blast waves but also due to an effect known as the **Parker instability** (Parker, 1966). The magnetic field cannot be in a stable equilibrium if the magnetic field lines are parallel to the Galactic plane. Instead, the field lines arch upward (see Figure 2.18) and thereby allow the interstellar gas to move downward along the field lines, under the influence of gravity. The high-energy cosmic rays can however travel along the field lines to reach large heights above the Galactic plane, from where we can detect their synchrotron emission. Figure 2.19 reveals Galactic synchrotron emission in the radio band.

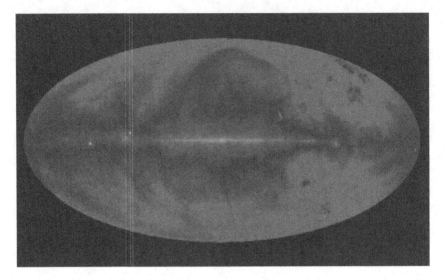

Figure 2.19 An all sky map of emission at 408 MHz, from data of Haslam et al. (1982), dominated by synchrotron emission from high-energy charged particles. Credit: C. Haslam, NASA.

Cosmic rays are thought to originate when charged particles get accelerated to high speeds by the fast-moving shock waves emerging from supernovae. They are detectable in the interstellar medium through their synchrotron emission as they spiral around the interstellar magnetic field.

A final intriguing thought is that the energy density of cosmic rays in the ISM is ~ 1 eV cm^{-3}, which is about the same as the energy density of kinetic (i.e., turbulent), magnetic, and radiation energy densities in the ISM. This equipartition implies a deep connection between the various processes of energy generation and transfer in the ISM.

2.4.5 Emission Nebulae (H II Regions)

In astronomical parlance, the Roman numeral I signifies a neutral species, II signifies a singly ionized species, III a doubly ionized species, and so on. H II regions are ionized gas nebulae consisting of mostly hydrogen, like all other gas in the universe, and surround massive stars that emit significant amounts of ultraviolet radiation. H II regions are some of the brightest objects in the Galaxy, and are observed through their optical and radio emission. They are bright enough in the radio band and there is sufficiently little extinction at those wavelengths that they can be observed clear across the breadth of the Galaxy. H II regions are therefore important signposts of star formation and spiral structure in our Galaxy. See Figure 2.20.

The recombination of protons and electrons generally leads to atomic hydrogen in a highly excited state, and the electron then cascades downward

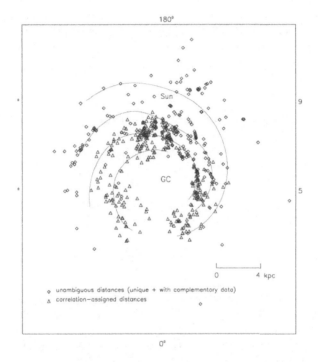

Figure 2.20 Positions of 550 H II regions measured by radio emission reveal Galactic spiral structure. The diamonds correspond to sources for which the distance is obtained directly, and the triangles represent sources for which the distance is obtained through a luminosity-diameter relation. The solid lines are a spiral arm model of the Galaxy by Taylor & Cordes (1993). Credit: Paladini et al. (2004), reproduced with permission from Oxford University Press (OUP).

through multiple transitions until it reaches the ground state. Along the way, there will be radio lines emitted, e.g., the transition $n : 93 \to 92$ that yields a photon with wavelength 3.61 cm and frequency 8.30 GHz. The H II regions can be seen in optical emission in the $3 \to 2$ Balmer α (known as Hα in astronomy) transition at 656 nm that gives them their distinctive red color.

To evaluate the size of emission nebulae, we apply a steady state condition, i.e., that the rate of ionization is the same as the rate of recombination. Let N_* be the number of Lyman continuum ($\lambda < 91.2$ nm) photons that leave the central star(s) per unit time. If R is the number of recombinations of electrons of number density n_e and ions of number density n_i into H atoms per unit volume per unit time, then

$$R = \alpha_r n_e n_i = \alpha n^2, \tag{2.6}$$

where α_r, with units of cm^3 s^{-1}, is the recombination coefficient, and we have used charge neutrality and that the gas is mostly hydrogen to argue that $n = n_e = n_i$.. The steady state condition is then

$$N_* = \frac{4}{3}\pi r_s^3 R, \tag{2.7}$$

which yields the equation for the Strömgren radius,

$$r_s = \left(\frac{3N_*}{4\pi\alpha n^2}\right)^{1/3}. \tag{2.8}$$

However, note that H II regions are rarely spherical, since the external medium need not be uniform. The radius of an H II region calculated here should also be thought of as just an initial value, as the interior of the H II region will reach an equilibrium temperature of approximately 10^4 K and become overpressured relative to its surroundings. This will lead to a dynamically expanding H II region, with a complex structure driven by the interplay of mechanical expansion and ionization structure, as well as inhomogeneities in the external medium.

2.5 BLAST WAVES

Supernovae and stellar winds are major contributors to the mass and energy transfer to the ISM. A supernova is a one-time deposition of a large amount of energy ($\approx 10^{51}$ erg) into the interstellar medium at the end of the life of a massive star. A stellar wind is a continuous flow of matter from a massive stars with speeds that can be in the thousands of km s^{-1} and mass loss rates $\approx 10^{-6}\ M_\odot$ yr^{-1}, thus depositing a comparable amount of energy as a supernova over the few Myr lifetimes of these stars. Here, we look at their effects in a simplified way and focusing on the case of a one-time energy deposition.

A single-point deposition of energy E_0 into a medium of density ρ_0 leads to a classical solution for an adiabatic blast wave. See Figure 2.21. A very elegant analytic solution is found in the limit of a very strong adiabatic shock wave, for which the external boundary conditions are unimportant. The instantaneous position $R_s(t)$ of the shock front can then only depend on E_0 and ρ_0. However, these two parameters cannot be combined to make a unit of length. In this situation (one where the number of parameters is less than the number of fundamental units), we can look for a so-called "self-similar" solution, by including the independent variable t in the dimensional analysis approach to finding an expression for $R_s(t)$. If we adopt an expression $R_s = \xi_0 E_0^\alpha \rho_0^\beta t^\gamma$, where ξ_0 is a dimensionless constant, and match the units on both sides of the equation, we obtain a set of equations that can be solved for α, β, and γ. This leads to a solution

$$R_s(t) = \xi_0 \left[\frac{E_0}{\rho_0} \right]^{1/5} t^{2/5}. \tag{2.9}$$

where ξ_0 is presumed to be of order unity. So, with very little effort invested, we find that the shock front expands with time in proportion to $t^{2/5}$. This remarkable prediction is borne out very well by observations of the shock front radius after a nuclear explosion in the Earth's atmosphere. We can see that the shock velocity $v_s(t) = dR_s(t)/dt$ scales as $t^{-3/5}$, or that

$$v_s(t) = \frac{2}{5} \frac{R_s}{t}. \tag{2.10}$$

Hence, observations of a size and velocity (estimated through Doppler shift) can yield a quick estimate of the age t of a remnant.

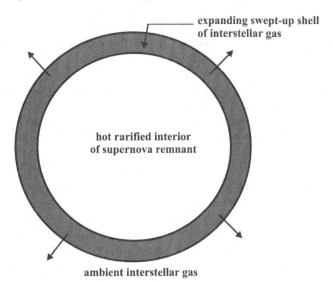

Figure 2.21   Illustration of a spherical supernova blast wave.

Eventually, a remnant will lose enough energy by radiation that energy will no longer be conserved. The swept-up shell of material will then enter the snowplow phase (momentum conservation). We can make a quick estimate of the motion in this phase as follows. If the ejected mass is

$$M_0 \approx 4M_\odot, \tag{2.11}$$

and

$$v_0 = 5000\,\mathrm{km\,s^{-1}}, \tag{2.12}$$

then the swept-up mass M $(= 4\pi r^3 \rho_0/3)$ when the shell velocity is v can be related to M_0 and v_0 by momentum conservation,

$$(M + M_0)v = M_0 v_0. \tag{2.13}$$

The shell dissipates when $v \sim 10\,\mathrm{km\,s^{-1}}$, which is the random speed of the ISM gas. Therefore, the amount of mass that is swept up is

$$\frac{M(v_0 - v)}{v} \approx \frac{Mv_0}{v} = 2000 M_\odot. \tag{2.14}$$

But what about the energy? We find that

$$\frac{E}{E_0} = \frac{(M + M_0)v^2}{M_0 v_0^2} = \frac{v}{v_0} = 2 \times 10^{-3}, \tag{2.15}$$

where we used the momentum conservation of Equation (2.13) to simplify the expression. So, where did all that energy go? The important point here is that momentum conservation does not imply energy conservation. Consider the analogy of two clay balls that collide and stick together, while conserving total momentum. Much of the energy has however disappeared into internal work to deform the initial object(s) and make them stick together. The work done to create the compressed shell can appear as thermal energy, but in the ISM, most of that energy is quickly converted to radiant energy and escapes the system. This keeps the shell from from heating up significantly. So, the answer to the question about missing energy is that most of it is lost to the system, due to the transparency to radiation generated in the compressed layer.

So far, we have considered an idealized spherical shell. However, it is in practice distorted for at least two reasons. If the external density is not uniform, and is decreasing rapidly, the blast wave may accelerate as it moves outward. In this case, in the frame of reference of an accelerating swept-up shell of matter, an effective gravity in the non-inertial frame is pointed back toward the source. Since there is also a lower density medium inside the shell, this leads to a Rayleigh–Taylor instability. The Crab Nebula clearly demonstrates this phenomenon, as seen in Figure 2.10.

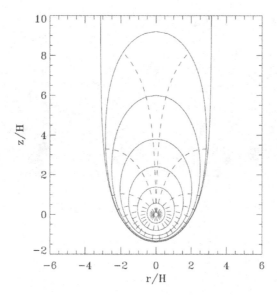

Figure 2.22 Blast wave propagation in a stratified atmosphere based on an analytic solution by Kompaneets (1960). The solid lines represent a sequence of shapes of a shock front as it propagates in a density profile of the form $\rho = \rho_0 \exp(-z/H)$. The dashed lines are streamlines of the flow. The final "U-shape" corresponds to blowout through the atmosphere. Credit: Basu et al. (1999), reproduced with permission from the AAS.

If the surrounding atmosphere is stratified, say in the vertical direction (perpendicular to the Galactic plane), then the blast wave will initially be spherical but will begin to elongate in the vertical direction once the radius of the shell becomes comparable to the scale height H of the vertical gas distribution (see Figure 2.22). Eventually, the blast wave may blow out the top, following the path of least external pressure resistance. Perhaps, the best way to see a blast wave that blows out through a galactic atmosphere is to observe the remnant in an external galaxy. Just such an observation was made by Cecil et al. (2001), revealing a large superbubble in the galaxy NGC 3079, and is shown in Figure 2.23.

Figure 2.23 A concave bowl representing a superbubble is seen in the galaxy NGC 3079. Emission from N II is in blue, emission in the $3 \rightarrow 2$ transition of hydrogen (Hα) is in red, and infrared band starlight is in green. Credit: Cecil et al. (2001), reproduced with permission from the AAS.

CHAPTER 3

Star Formation

3.1 INTRODUCTION

Star formation is the process of transformation of diffuse interstellar gas into collapsed objects. Stars can be defined as objects that eventually reach a state of balance between their inward self-gravitational force and the outward force due to thermal pressure. That thermal pressure is maintained through energy released in sustained thermonuclear fusion. We now know that substellar objects known as **brown dwarfs** may also form in a similar manner, through direct gravitational collapse. Another important class of object are planets, which are clearly very important to our existence. They are thought to form within circumstellar disks that exist around young stars.

What then are the scientific definitions of planets, brown dwarfs, and stars? We can adopt the following criteria based on the presence or absence of nuclear fusion within an object. A planet is a gravitationally bound object in which nuclear fusion is never initiated. Theoretical calculations show that the upper mass limit for no nuclear fusion at all is about 13 $M_{\rm Jup}$, where $M_{\rm Jup} = 1.90 \times 10^{30}$ g is the mass of the planet Jupiter. Above this mass nuclear reactions may occur but cannot achieve a steady state balance between energy generation in the interior and energy losses by radiation at the surface. These objects are brown dwarfs and have masses M in the approximate range 13 $M_{\rm Jup} \lesssim M \lesssim$ 80 $M_{\rm Jup}$. Above about 80 $M_{\rm Jup}$ (0.075 $M_{\odot}$) the hydrogen fusion is sufficient to generate enough luminosity to balance surface losses, and the object settles into a steady-state that is known as the **main sequence**. More discussion of that phase can be found in the Chapters on Stars and Stellar Evolution.

The above definitions are based on masses and the consequent level of nuclear fusion, but a different way of looking at the objects is through the formation mechanisms. Stars are expected to form through direct collapse, but many stars come in binary or multiple systems. In these cases, the secondary star(s) that accompany the primary star may have come from gravitational instability within the disk, leading to a collapsing clump in the disk. This process is also known as disk fragmentation. Brown dwarfs may come from either or both processes of direct collapse or disk fragmentation. On the other

DOI: 10.1201/9781003215943-3

hand, planets were traditionally thought to come only from the bottom-up process of the coagulation of solids in a circumstellar disk. However, there is more recently the idea that at least some of the massive gas giant planets may have formed by disk fragmentation. Figure 3.1 illustrates the various mechanisms.

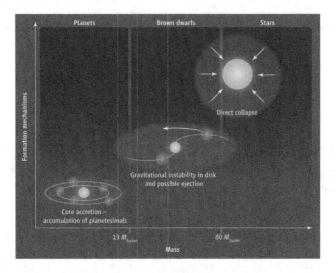

Figure 3.1 Different mechanisms that can lead to the formation of stars, brown dwarfs, and planets. Credit: Basu (2012), reproduced with permission from the American Association for the Advancement of Science.

3.2 CLOUDS AND COLLAPSE

3.2.1 Molecular Cloud Fragmentation

The coldest objects in the universe are the molecular clouds, with temperatures in the range 10–50 K. Molecular transitions offer numerous pathways for emission that are also very effective in cooling the cloud. Some of the most effective pathways are the rotational transitions of the CO molecule. The rules of quantum mechanics govern the allowed rotational energy levels of a diatomic molecule, and only levels described by integer quantum values J are allowed. These levels have energy

$$E = \left(\frac{h}{2\pi}\right)^2 \frac{J(J+1)}{2\mu r^2}, \tag{3.1}$$

where h is Planck's constant, $\mu = m_1 m_2/(m_1 + m_2)$ is the reduced mass of the two atoms of mass m_1 and m_2, and r is the mean separation (bond length) between the atoms. The $J = 1 \to 0$ transition of CO corresponds to a wavelength 2.6 mm (frequency 115 GHz) and is a major coolant of molecular clouds as well as a commonly observed spectral line for molecular cloud observations. Although the most abundant molecule by far is molecular hydrogen, H_2, it does not have transitions in the wavelength range visible from the Earth's surface. Hence, our observational knowledge of the structure of molecular clouds comes mainly from measuring the emission from trace molecules like CO. These surveys of CO emission show that all star formation in our Galaxy occurs within molecular clouds.

Figure 3.2 shows a map of ^{13}CO emission from the Taurus Molecular Cloud, located at a distance of about 140 pc, and one of the nearest regions of ongoing star formation. The ^{13}CO is an isotopologue of normal CO (i.e., ^{12}CO), meaning that it differs from CO in its isotopic composition. This species is particularly useful to measure the column density of gas, as the low abundance of the ^{13}CO species means that ^{13}CO emission from all layers of the molecular cloud should be reaching us, with little self-absorption by other ^{13}CO molecules along the line of sight. Astronomers call such an emission line **optically thin**. Stars tend to form within dense subregions (known as **cores**) of the cloud that have a typical size scale ~ 0.1 pc.

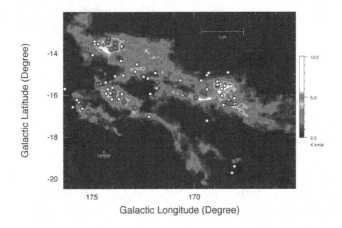

Figure 3.2 Map of ^{13}CO $J = 1 \to 0$ emission at 110 GHz from the Taurus Molecular Cloud, using data from Onishi et al. (1996). If observable by the naked eye, this region would be more than 30 Moon diameters across in the sky. The locations of protostars are marked in red and those of pre-main-sequence stars in white. Credit: T. Onishi.

A map of the same region made by the Planck satellite, in which the colors represent column density of gas and the drapery pattern follows the local direction of the magnetic field inferred from polarized emission, shows that the magnetic field is oriented roughly perpendicular to the dense regions where stars are forming. See Figure 3.3. The Taurus star-forming region may be a collection of ribbon-like structures in which the shortest dimension is aligned with the large-scale magnetic field direction. Gas can settle most efficiently along the direction of the magnetic field, since magnetic fields only exert forces perpendicular to their local direction.

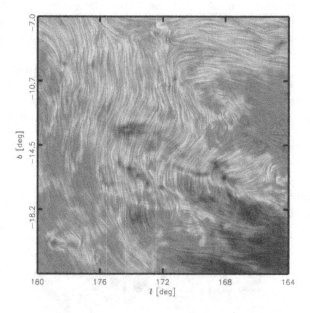

Figure 3.3 The magnetic field and gas column density measured by the Planck satellite toward the Taurus Molecular Cloud. The colors represent the column density and the drapery pattern indicates the orientation of the magnetic field lines, which is inferred to be orthogonal to the orientation of submillimeter polarization of dust emission. Credit: Planck Collaboration et al. (2016), reproduced with permission from the European Southern Observatory (ESO).

3.2.2 Jeans Instability Criterion

An important question is: what is the minimum gas mass that has sufficiently strong gravity to essentially collapse due to its own weight? The answer was found by James Jeans (1877–1946) in 1902 (Jeans, 1902), and we derive it here in a simplified manner. For gravity to overcome internal thermal support, the (negative) gravitational energy E_{grav} must have greater magnitude than the (positive) thermal energy E_{therm}. A more rigorous approach of directly integrating the force balance equation for a sphere shows that the criterion is

$$2E_{therm} < -E_{grav} . \tag{3.2}$$

Assuming a sphere of uniform density ρ, the mass M and radius R are related by $M = (4/3)\pi R^3 \rho$, and $E_{grav} = -(3/5)GM^2/R$. At a temperature T, the gas particles have an average thermal energy $(3/2)kT$. Hence, the total thermal energy is $E_{\text{therm}} = (3/2)NkT$, where $N = M/(\mu m_H)$ is the total number of particles, m_H is the mass of a hydrogen atom, and μ is the multiplicative factor such that $m = \mu m_H$ is the mean particle mass. Atomic hydrogen gas clouds with 10% helium number content have $\mu = 1.3$, whereas molecular hydrogen clouds have $\mu = 2.3$.

Combining the above relations, Equation (4.10) can be written as

$$M > M_J \equiv k_1 \frac{c_s^3}{G^{3/2}\rho^{1/2}}, \qquad (3.3)$$

where M_J is the Jeans mass, $c_s = (kT/\mu m_H)^{1/2}$ is the isothermal sound speed, and we have used $M = (4/3)\pi R^3 \rho$. The leading dimensionless constant is $k_1 = (375/4\pi)^{1/2}$ in this derivation. One can also write the criterion as

$$R > R_J \equiv k_2 \frac{c_s}{(G\rho)^{1/2}}, \qquad (3.4)$$

where R_J is the Jeans length and $k_2 = (3k_1/4\pi)^{1/3}$. More exact calculations lead to modified values of k_1 and k_2, but they are always order unity constants.

3.2.3 Magnetic Field Support

In the cgs system, the magnetic energy density is $B^2/8\pi$, where B is the magnetic field strength. The total magnetic energy in a volume V is then

$$E_{\text{mag}} = \frac{B^2}{8\pi}V. \qquad (3.5)$$

In a situation where the magnetic field is well coupled to the plasma, i.e., magnetic flux freezing applies, gravity wins, and collapse happens only if $E_{\text{mag}} < -E_{\text{grav}}$. Simplifying to the case of uniform magnetic field B, this leads to

$$\frac{B^2}{8\pi}\frac{4}{3}\pi R^3 < \frac{3}{5}\frac{GM^2}{R}. \qquad (3.6)$$

Here, for simplicity, we ignore the additional support of thermal pressure. Letting $\Phi = B\pi R^2$ be the magnetic flux, this criterion becomes

$$\frac{M}{\Phi} > \left(\frac{M}{\Phi}\right)_{\text{crit}} \equiv \frac{c_1}{G^{1/2}}. \qquad (3.7)$$

Here, $(M/\Phi)_{\text{crit}}$ is the critical mass-to-flux ratio and $c_1 = (5/18\pi^2)^{1/2}$. More exact calculations can be performed (Mouschovias & Spitzer, 1976), but the value of the dimensionless constant c_1 remains comparable to the value derived here. Since we ignored the thermal pressure in the above derivation, this is really a necessary condition for gravitational collapse, but a sufficient criterion is that a gas cloud also has $M > M_J$. Interstellar gas clouds taken as a whole tend to have $M \gg M_J$; hence, we usually apply Equation (3.7) directly.

3.2.4 Gravitational Collapse

So what happens when a gas clump exceeds the Jeans criterion and begins to collapse? Prior to the late 1960s, many astrophysicists believed that gravitational collapse and star formation occurred in a **homologous** manner, e.g., it could be thought of as the collapse of a uniform density sphere that kept a spatially uniform density profile but with the density increasing systematically with time. Numerical simulations were performed in the late 1960s on the collapse of gas that remains largely isothermal at low densities due to its ability to radiate away the internal energy generated by compression (Bodenheimer & Sweigart, 1968; Larson, 1969; Penston, 1969). The results showed, surprisingly at the time, that the collapse is highly **nonhomologous**, exhibiting the runaway collapse of a small central region.

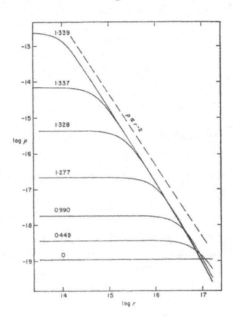

Figure 3.4 A time sequence of profiles of density ρ versus radius r in a collapsing sphere. The curves are labeled with the times in units of 10^{13} s $\simeq 3.2 \times 10^5$ yr since the beginning of the collapse. The asymptotic profile is one in which $\rho \propto r^{-2}$. Credit: Larson (1969), reproduced with permission from OUP.

Figure 3.4 shows the development of runaway collapse of an isothermal sphere of molecular gas (mostly H_2) that is initially uniform density and is unstable according to the Jeans criterion. This is a spherically symmetric radial infall model with no rotation or magnetic effects. The striking result is the runaway collapse of a shrinking central region beyond which is the development of a power-law profile in density $\rho \propto r^{-2}$. Once the central region reaches a density that is high enough for photon trapping, it is opaque, and the temperature rises until the collapse is halted inside this small central region. This process starts at a number density $n \approx 10^{10}$ cm^{-3} and the resulting object is known as the **first hydrostatic core**, or **first core** for short. Once the first core reaches a temperature of about 2000 K, the hydrogen molecules (H$_2$) are dissociated. The energy absorbed in the dissociation process reduces the thermal support in the core and it undergoes a second collapse phase and contracts down to stellar dimensions ($\sim$ few $R_\odot$). This stellar core initially has a mass of about $10^{-3}\,M_\odot$ and the rest of the star formation can be thought of as an accretion process in which the remaining gas falls onto the stellar core or is diverted into a disk, jet, or outflow.

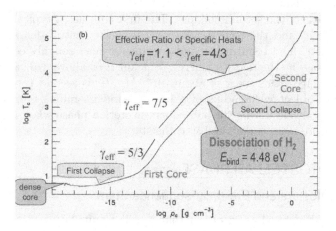

Figure 3.5 The evolution of temperature versus density in a star-forming gas cloud. Each of the phases in the temperature evolution is characterized by a distinct value of the effective ratio of specific heats, γ_{eff}. Adapted from Masunaga & Inutsuka (2000). Credit: S-i. Inutsuka.

Figure 3.5 shows the evolution of central temperature versus density in a collapsing star-forming cloud. The calculation takes into account various molecular cooling processes, emission by dust, and radiative trapping and re-emission of photons. The key element in analyzing this plot is the value of γ_{eff}, which is the exponent in the relation between pressure P and density ρ, i.e., $P \propto \rho^{\gamma_{\mathrm{eff}}}$. The ideal gas law $P = \rho k T / m$ then implies that $T \propto \rho^{\gamma_{\mathrm{eff}}-1}$. In the early collapse, the gas is transparent and the energy gained from compression is rapidly radiated away, causing an isothermal evolution. Once a number density of $n \approx 10^{10}$ cm^{-3} is reached, the opacity of the gas is sufficient that radiative trapping causes the temperature to increase gradually above the low value (~ 10 K) found in molecular clouds. At first, the slope of the temperature curve corresponds to $\gamma_{\mathrm{eff}} = 5/3$ ($T \propto \rho^{2/3}$ for $10\,\mathrm{K} < T < 10^2$ K). This seeming monatomic gas behavior is due to the fact that the rotational degree of freedom of H_2 is not excited in this low temperature regime, given that the minimum possible rotational energy gap is $\Delta E(J = 2 \to 0)/k = 512$ K (the rules of quantum mechanics allow only rotation level jumps of $\Delta J = \pm 2$ for H_2 since it does not have a permanent dipole moment). When the temperature rises higher, γ_{eff} becomes 7/5, characteristic of diatomic molecules. In both the cases, γ_{eff} is greater than the critical value of 4/3 for gas pressure support against self-gravity. Thus, the collapse is rapidly decelerated and forms a shock at the surface of the newly formed first core. Its radius is in the range of $1-10$ AU, and it consists mainly of H_2. As the accretion on to the first core continues and the temperature inside the first core rises above $\sim 10^3$ K, the dissociation of H_2, with binding energy 4.5 eV, begins. The dissociation of H_2 acts as an efficient coolant of the gas, which reduces γ_{eff} below the critical value of 4/3, and triggers the second collapse. The first core has a typical lifetime of only $\sim 10^3$ yr, and the second collapse causes the central density to reach a stellar value. The resulting second core has a mass roughly equivalent to the Jeans mass at the density where γ_{eff} again rises above 4/3, when all the H_2 is dissociated; this mass is about $10^{-3} M_\odot$. This second core is truly a stellar-sized object with radius of a few $R_\odot$ and its radius remains of this order of magnitude during the subsequent accretion phase when much of the remaining core mass is gathered into the star.

3.3 PROTOSTELLAR PHASE

3.3.1 Protostars and Pre-main-sequence Stars

A protostar is defined as an object that is still accumulating matter from its surroundings. Protostars are usually obscured from view by a layer of dust in the infalling envelope that surrounds them, so that much of what is observed is the infrared emission from the surrounding dust. The dust is heated by radiation emanating from an **accretion shock** created when infalling matter encounters the protostar's surface. The accretion shock itself may also be detectable through an excess ultraviolet emission.

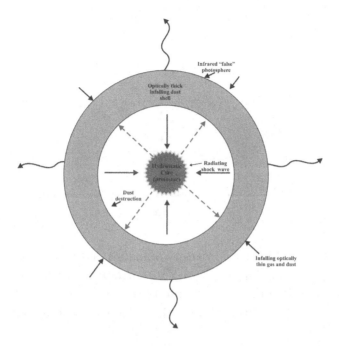

Figure 3.6   An illustration of an idealized spherically symmetric protostellar environment.

Figure 3.6 illustrates the main features of a model for protostar emission that is based on a spherical collapse model. Matter that reaches the protostar surface will form a shock front that radiates mainly at ultraviolet and optical wavelengths. However, the surrounding gaseous envelope (which is likely infalling) is optically thick at these wavelengths. The dust provides most of the opacity but is destroyed by ultraviolet radiation within an inner region where the temperature exceeds ≈ 1500 K. Beyond this **dust destruction front**, the dust creates an optically thick shell that absorbs the optical radiation and reradiates it in the mid- to far-infrared wavelengths. Observations show far-infrared emission similar to that predicted by the spherical models but much more emission at shorter wavelengths. This can be understood through the presence of rotation, which creates a disk-like configuration of gas and dust around the protostar. This configuration allows some of the emission from the hot central regions to reach the observer. Overall, the spectra are qualitatively of the form shown in Figure 3.7.

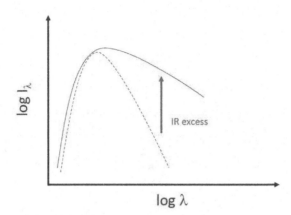

Figure 3.7 Illustration of the intensity of radiation versus wavelength (the spectral energy distribution) for a YSO surrounded by a disk. The presence of a disk is inferred from the infrared excess emission. This excess at long wavelengths significantly exceeds the expected blackbody emission (dashed line) from the surface of the YSO.

A pre-main-sequence (PMS) star is an object that has ended its main mass accumulation phase, although residual mass accretion from a surrounding disk may still be active. These objects have not reached a high enough central temperature to initiate hydrogen fusion; however, they are slowly contracting while releasing gravitational potential energy. In this way, the PMS stars are doing exactly what the 19th century physicists imagined that all stars were doing.

Observations of protostars and PMS stars are placed into four categories. These categories are based on features in their observed spectral energy distribution (SED), often characterized by significant infrared (IR) emission at $\sim 1\,\mu m$ and longer wavelengths from the surrounding nebulosity or disk. They are as follows.

Class 0: The SED peaks at $\sim 100\,\mu m$, with no emission detected short of $\sim 10\,\mu m$.

Class I: The SED is approximately flat or rising into the mid-IR.

Class II: The SED is decreasing into the mid-IR.

Class III: The SED shows little or no excess in the IR.

This classification is interpreted in terms of an evolutionary sequence. A Class 0 object is a protostar that is still accumulating a significant portion of its final mass (Andre et al., 1993), and is heavily obscured by the surrounding material, leading to emission that can peak at very long wavelengths $> 100\,\mu m$. The Class I to Class III objects have successively decreasing amounts of surrounding material (Adams et al., 1987). Class I objects still have some surrounding nebulosity in addition to a surrounding disk. Class II objects (otherwise known as T Tauri stars) are composed of a young PMS star and a circumstellar disk. Class III objects (otherwise known as weak-lined T Tauri stars) do not have a detectable circumstellar disk since it is presumably lost due to accretion and dispersal. The term young stellar object (YSO) is used to collectively refer to any of these objects, whether a protostar or a PMS star.

3.3.2 Outflows and Jets

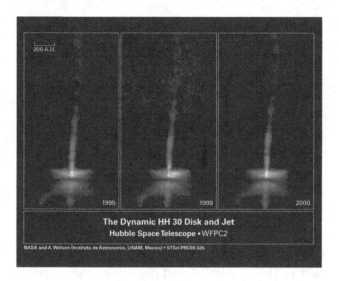

Figure 3.8 The object known as Herbig-Haro 30 (HH 30), a prototypical young star surrounded by a disk and jet. These *Hubble Space Telescope* images show changes over only a five-year period in the disk and jets, which are visible in optical emission while the young star is obscured by dust in the disk. Credit: NASA and STScI.

Two of the most prominent features of YSOs are the spectacular jets and outflows that emanate from their poles or nearby environment. See Figure 3.8. These bipolar flows were first detected in the 1980s, and were a surprise to astronomers. The jets are high-speed collimated material that emerge along the poles of a YSO. The typical speeds are of order 100 km s^{-1} and these are detected through optical emission from shocks and other interactions as the ejected material interacts with surrounding nebulosity. They are often referred to as **Herbig-Haro objects** and may extend up to pc scales. The molecular outflows are detected instead through mm emission from entrained molecules and are slower and have a much broader opening angle than the jets. The speeds are of order 10 km s^{-1}.

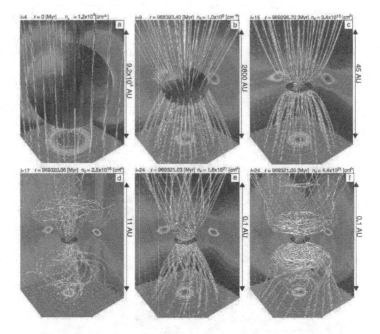

Figure 3.9 A sequence of images from a numerical simulation of the collapse of a rotating dense cloud core with an embedded magnetic field. The calculation is done on a nested grid, with grid levels denoted by the letter l and higher values corresponding to more inner regions. The elapsed time t in the simulation and density at the center, n_c, are also denoted in each panel. The red region is the densest region where the density is greater than $0.1\,n_c$. The magnetic field lines are represented by black-and-white streamlines. The density contours (false color and contour lines) and velocity vectors (arrows) on the cross sections in the $x = 0$, $y = 0$, and $z = 0$ planes are projected on the respective sidewalls of the images. Credit: Machida et al. (2007), reproduced with permission from the AAS.

Theoretical work has established that the jets and outflows are most likely driven by magnetic forces associated with the extreme pinching and twisting of the magnetic field lines in the vicinity of the deeply collapsed YSO. Such configurations of the magnetic field are implied by even partial magnetic flux freezing during the collapse process. Figure 3.9 shows an evolutionary sequence of a gravitational collapse simulation with an embedded magnetic field (Machida et al., 2007). The collapse leads to an increasingly pinched magnetic field configuration, especially after the formation of a central protostar (panels c, d, e, f). The flared-out field lines in a surrounding rapidly rotating disk that is associated with the first core can fling material outward along the magnetic field lines, in a phenomenon known as the **magnetocentrifugal effect**. In this region, the three components of the magnetic field in a cylindrical coordinate system are comparable to each other: $B_r \approx B_z \approx B_\phi$. Closer to the protostar, the rotational energy dominates the magnetic energy. The magnetic field is twisted into a toroidal shape, and the hierarchy of field components is $B_r \ll B_z \ll B_\phi$. The vertical gradient of the toroidal magnetic field drives high speed jets along the magnetic poles, which are parallel to the rotation axis. This phenomenon is known as a **magnetic tower flow**. Figure 3.10 illustrates the mechanisms driving the outflow and the jet.

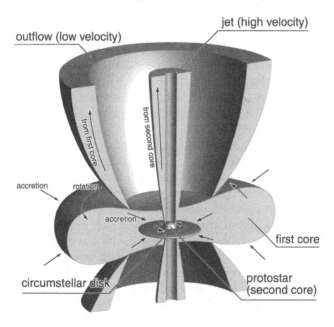

Figure 3.10 Illustration of the driving mechanisms for a molecular outflow and a jet. Credit: Machida et al. (2008), reproduced with permission from the AAS.

3.3.3 Disk Evolution

One of the most striking aspects of star formation is the formation of a protostellar disk, sometimes also known as a circumstellar disk. Technically, a protostellar disk is a disk that surrounds a protostar (an object that is still accumulating a significant portion of its final mass, and is shrouded at least partially by surrounding gas and dust) and a circumstellar disk is one that surrounds a PMS star or even a later stage star. Here, we use the two terms interchangeably.

A disk is expected to form around a forming star simply due to the concept of angular momentum conservation. Although magnetic fields and outflows reduce the amount of angular momentum in the infalling gas, we can make a quick estimate for the size of a disk from simple mechanical considerations. If a dense core collapses due to its own gravity with a mass M and specific angular momentum j $(= \Omega r^2$ where Ω is the instantaneous rotation rate at a radius r) in a mass shell at the edge of the core, then that mass shell will reach a centrifugal balance in its rotating frame when it has fallen inward to a centrifugal radius r_c such that

$$\frac{j^2}{r_c^3} = \frac{GM}{r_c^2} \Rightarrow r_c = \frac{j^2}{GM}. \tag{3.8}$$

Since j can also be estimated from the initial conditions of collapse, we use typical observed values for dense cores in molecular clouds, $\Omega_{core} \approx 10^{-14}$ rad s^{-1} and $r_{core} \approx 0.1\,\mathrm{pc} = 3 \times 10^{17}$ cm to find $j = \Omega_{core} r_{core}^2 \approx 10^{21}$ cm^2 s^{-1}. For a collapsing object of mass $M = 1\,M_\odot$, this leads to $r_c \approx 500$ AU. Most observed disks are within this size range, with a typical observed average value at late stage evolution of ~ 100 AU.

Theoretical calculations of the formation of circumstellar disks need to account for multiple physical effects, as illustrated in Figure 3.11. A central YSO is surrounded by a disk that accretes matter from a collapsing cloud core. The infalling material lands on the outer edge of the disk and is subsequently transported inward by torques due to viscosity and the gravity of nonaxisymmetric structures. The star also radiates photons, and part of this stellar irradiation is absorbed by the flaring disk surface and heats the disk interior. Another source of external heating is the background irradiation from the natal molecular cloud. The thermal energy generated in the disk interior by viscosity and possible shock dissipation is transported to the disk surface by radiation. This energy escapes the disk through surface cooling by radiation.

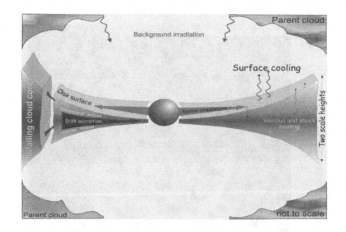

Figure 3.11 An illustration of a central protostar and surrounding disk. See the text for a detailed explanation. Credit: Vorobyov & Basu (2010), reproduced with permission from the AAS.

Numerical simulations of disk evolution reveal a complicated time-dependent evolution. Disks form due to reasons of a centrifugal barrier as described above, but their subsequent evolution reveals much nonlinear and episodic events. In the first ~ 0.5 Myr of evolution there is sufficient mass infall from the surrounding cloud core to send the disk into recurring events of gravitational instability. The instability occurs when the value of the dimensionless parameter

$$Q = \frac{c_s \Omega}{\pi G \Sigma} \tag{3.9}$$

falls below a value of $1/2$ (Toomre, 1964). Here c_s is the isothermal sound speed, Ω is the rotation rate, and Σ is the mass column density. The ongoing mass infall drives up the column density and therefore drives down the value of Q toward the unstable regime. When $Q < 1/2$, nonlinear spiral arms and clumps form within the disk as shown in a simulation snapshot in Figure 3.12.

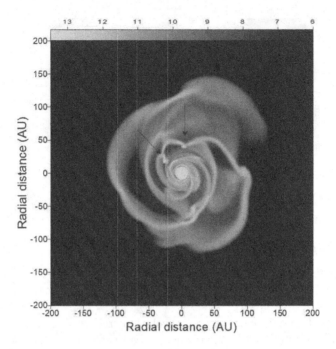

Figure 3.12 A snapshot image of a protostellar disk from a time-dependent computer simulation. The gas volume density is shown with the scale bar in logarithmic values with units of cm^{-3}. The snapshot is from a time when nonlinear gravitational instability has set in. The proto-stellar/protoplanetary embryos, identified as regions with number density $n > 10^{13}$ cm^{-3}, are indicated with arrows. The bright circle in the center represents the protostar plus some circumstellar matter. Credit: Vorobyov & Basu (2006), reproduced with permission from the AAS.

In the mid-2000s, Vorobyov & Basu (2006) found an interesting result: the clumps feel a strong gravitational torque from the spiral arms and are driven inward toward the YSO. The arrival of a clump at the YSO leads to a burst of mass accretion as well as a significant outburst of luminosity. The released gravitational potential energy when converted largely to radiation yields an accretion luminosity

$$L_{\text{acc}} \approx \frac{GM}{R}\frac{dM}{dt}, \qquad (3.10)$$

where M and R are the mass and radius of the protostar and dM/dt is the accretion rate of matter to the YSO. Figure 3.13 shows the evolution of episodic accretion over approximately the first 0.5 Myr of disk evolution. Strong and frequent episodic bursts at early times give way to a more quiescent accretion rate by the end of the time period. The rapid rise of mass accretion and luminosity at early times signify the formation of a protostar from the collapsing cloud core.

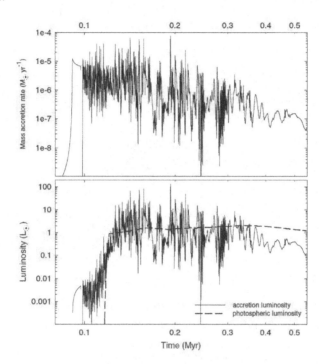

Figure 3.13 Mass accretion rate onto a star (top) and stellar luminosity (bottom) as a function of time elapsed since the beginning of collapse in a numerical simulation of star formation. In the bottom panel the solid line shows the accretion luminosity and the dashed line shows the estimated luminosity from the photosphere of the central YSO. Credit: Vorobyov & Basu (2010), reproduced with permission from the AAS.

The occurrence of gravitational instability at early times in the disk evolution and the typical clump mass of $\sim 10\,M_{\rm Jup}$ hints at the possibility that giant planets may form by direct collapse of gas fragments in circumstellar disks. In some simulations, the clumps carve out a gap in the gas disk and thereby minimize any gravitational torques that may drive them further inward, thereby allowing for a stable orbit. This method of forming a giant planet is an alternative to the standard core accretion model of planet formation (Safronov, 1972), in which the long-term coagulation of dust grains in the disk leads to larger objects known as planetesimals, and the formation of planetary cores that are built by the accretion of planetesimals. As the cores grow, their ability to accrete gas from the surrounding disk increases. When the core is sufficiently massive, rapid gas accretion occurs onto the core and a gas giant planet is formed. Less massive cores lead to the formation of terrestrial planets, which also gather some gas but later lose most of it due to their weak gravity. Figure 3.14 illustrates this model.

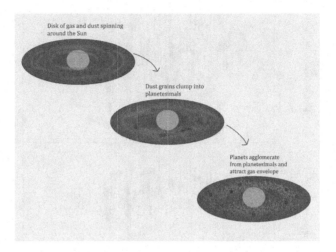

Figure 3.14 An illustration of the sequence of events leading to planet formation in the standard core accretion model.

Observations of circumstellar disks have been revolutionized by the *Atacama Large Millimeter/Submillimeter Array (ALMA)*. It measures the emission from dust with unprecedented resolution and sensitivity, and is able to detect the faint mm emission from disks with a resolution of 10 AU or less for the nearest star-forming regions. Figure 3.15 shows a compendium of images of disks in nearby star-forming regions (Andrews et al., 2018). The remarkable result emerging from these observations is that there is an incredible variety of substructures in these disks: gaps, rings, and spirals are found in most of them. These structures are seen in very young disks, less than 1 Myr old in some cases, and argue for a more rapid period of planet formation than in conventional core accretion models.

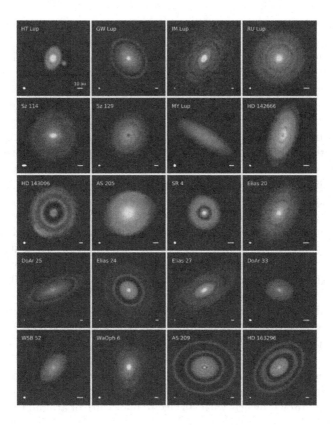

Figure 3.15 Gallery of circumstellar disk images in 1.25 mm continuum emission measured by the ALMA telescope as part of the *Disk Substructures at High Angular Resolution Project (DSHARP)*. Beam sizes and 10 AU scalebars are shown in the lower left and right corners of each panel, respectively. All images are shown with a stretch factor in intensity so as to accentuate fainter details without over-saturating the bright emission peaks. Credit: Andrews et al. (2018), reproduced with permission from the AAS.

3.4   STELLAR MASSES

3.4.1   The Substellar Mass Limit

In the early 1960s, Shiv Kumar (Kumar, 1963) and Hayashi and Nakano (Hayashi & Nakano, 1963) showed independently that a gravitationally-bound object could, if its mass is low enough, reach a state of hydrostatic equilibrium in which the internal pressure due to the zero-point motions inherent in quantum mechanics (known as **degeneracy pressure**) was enough to balance the weight of gravity. In this case, the object would never contract to a state in which nuclear fusion could set in at a high enough rate to balance the surface energy loss by radiation. An object that does achieve such an energy balance is a main-sequence star. An object with insufficient nuclear fusion to achieve energy balance is a brown dwarf, although it is never really brown in color. The **substellar mass limit**, below which an object is identified as a brown dwarf, is currently estimated to be $\approx 0.075\,M_{\odot}$ (Chabrier & Baraffe, 2000). Figure 3.16 shows how the central temperature T_c of objects of mass near the substellar limit varies as their radius changes. Each point on a line represents a different equilibrium state. As the radius decreases from large values at the right, the equilibria achieve increasing values of T_c. However, past a certain amount of compression, T_c declines again, as the degeneracy pressure becomes significant, thus requiring less thermal pressure to achieve an equilibrium state. If the peak value of T_c remains below the value for initiating hydrogen fusion, which is at least several times 10^6 K, or just remains low enough to lead to insufficient nuclear fusion to balance the surface energy loss, then the object becomes a brown dwarf. A brown dwarf will gradually cool and become less luminous over time.

Observationally, the spectral classes L, T, and Y cover the observed range of brown dwarf surface temperatures. Objects in class L have surface temperatures in the range of 1300 K to 2000 K, and are typically a red-brown color. Class T objects are even cooler, with surface temperatures in the range 700 K to 1300 K, and have a dark magenta color. The class Y objects are extremely faint and only a handful of candidate objects have been detected so far. These are ultra-cool brown dwarfs with surface temperature of around 600 K or less.

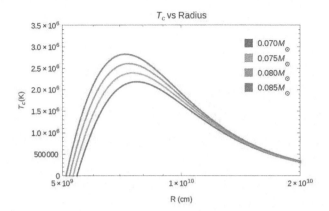

Figure 3.16 The variation of central temperature T_c versus radius R for equilibrium states of gravitationally bound objects. Each colored line represents the sequence of equilibrium states for an object of a fixed mass as labeled. Due to the onset of degeneracy pressure at high density, there is a maximum attainable temperature for each mass. For an object at the substellar mass limit, the peak temperature is just high enough for hydrogen fusion. Based on data presented in Auddy et al. (2016). Credit: S. Auddy.

3.4.2 Initial Mass Munction

Stars come in a wide range of masses, with a minimum mass $\approx 0.075\,M_\odot$ and an unknown maximum mass that is estimated to be $\approx 100\,M_\odot$. The minimum mass is associated with the onset of sustained hydrogen fusion whose energy output can balance surface energy losses by radiation. However, the star formation process is unaware of this fusion constraint and we now know that many objects are formed with a mass below the substellar mass limit. The upper limit is set by radiation pressure, which can in principle overcome the gravitational pressure for very massive stars; this limit is not exactly known but is estimated to be $\approx 100\,M_\odot$ under normal circumstances of star formation in our Galaxy.

Since stars of different masses have different lifetimes, may suffer intense mass loss at some point in their lives, or even end their lives destructively in a supernova, an important concept is the **initial mass function** (IMF), or the distribution of masses for newly born stars (or brown dwarfs). The best way to measure this is to look for young stellar clusters, where all the objects would have formed at about the same time and have the same chemical composition.

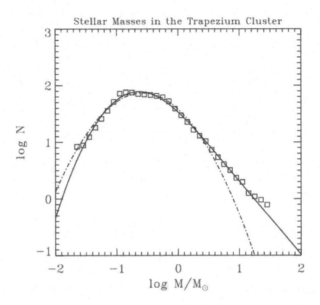

Figure 3.17 Histogram of stellar masses in the Trapezium cluster of the Orion Nebula. The squares are mass bins from the observations of Muench et al. (2002), the dash-dotted line is a best-fit lognormal distribution and the solid line is a best-fit modified lognormal power-law (MLP) distribution that is described by Basu et al. (2015).

A very well-studied stellar cluster is the Orion Trapezium Cluster, located in the constellation Orion and at a relatively close distance of about 410 pc. Figure 3.17 shows a compilation of the distribution of estimated masses of objects in the cluster (Muench et al., 2002) and a best-fit theoretical model. The masses are estimated from the observed luminosities, using theoretical models of the evolution of pre-main-sequence stars and brown dwarfs. The data shows a broad peak at about $0.2\,M_\odot$ and, for $M > 1M_\odot$, a straight line in the log–log plot, otherwise known as a power-law profile. Since each box height represents the logarithm of the number ΔN of objects contained in each mass bin of width $\Delta \log M$, a straight line of slope $-\alpha$ implies that the distributon function of objects satisfies $\Delta N/(\Delta \log M) \propto M^{-\alpha}$. This compilation of the IMF is very similar to that found from compilations of stars across the Galaxy, which also yield $1 < \alpha < 2$ for the mass range $M > 1M_\odot$. The overlaid solid line is an analytic function that transitions from a peaked distribution at low masses to a power-law function at intermediate and high masses (Basu et al., 2015).

CHAPTER 4

Stars

4.1   INTRODUCTION

Stars have fascinated humankind since the dawn of our species. On a clear night and away from man-made lights, you can see hundreds of stars and perhaps a few bright planets as well. We now know that stars are the same kind of object as our Sun but located at immense distances from us. If the Sun were a grain of sand, the nearest star would be 6 km away! The stars also come in versions that can be much more or much less massive than our Sun, and their luminosities vary by a staggering amount. The brightest star may be a million times brighter than our Sun, and the dimmest may be one ten thousandth of our Sun in brightness.

In this Chapter, we review some properties of gravitation that apply to stars, and then cover many aspects of the observation of stars.

4.2   STELLAR STRUCTURE

4.2.1   Gravitation

Stars are strongly self-gravitating and the gravitational field at the surface of a spherical star of mass M and radius R is

$$g = -\frac{GM}{R^2}. \tag{4.1}$$

Will gravity lead to collapse? To investigate, we can look at two limiting cases.
Free fall:
For a spherically symmetric collapse, the instantaeous position of a mass shell $r(t)$ that encloses a mass M can be calculated from Newton's second law,

$$\ddot{r} = -\frac{GM}{r^2}. \tag{4.2}$$

Hydrostatic equilibrium:
In this case, every spherical mass shell of thickness dr will feel an inward gravitational force $-Gm(r)dm(r)/r^2$ where $m(r)$ is the enclosed mass,

DOI: 10.1201/9781003215943-4

$dm(r) = 4\pi r^2 \rho(r) dr$ is the mass in the shell, and $\rho(r)$ is the local density. The inward force must be balanced by a net outward force due to the internal pressure, which has magnitude $4\pi r^2 (P(r) - P(r+dr))$. Balancing these forces and realizing that $dP/dr \equiv (P(r+dr) - P(r))/dr$, we find the force balance equation

$$\frac{dP}{dr} = -\rho(r) \frac{Gm(r)}{r^2}. \tag{4.3}$$

We study some aspects of each of these limits in turn.

4.2.2 Free Fall

We can multiply both sides of Equation (4.2) by $\dot{r}$ and realize that $\dot{r}\ddot{r} = \frac{d}{dt}(\frac{1}{2}\dot{r}^2)$. This allows us to integrate both sides of the equation, and using the initial condition of starting from rest at a radius r_0, we find

$$\frac{1}{2}\left(\frac{dr}{dt}\right)^2 = GM\left(\frac{1}{r} - \frac{1}{r_0}\right). \tag{4.4}$$

The time required for a mass shell to reach $r = 0$ is the **free-fall time**. This can be expressed as

$$t_{\text{ff}} = \int_0^{t_{\text{ff}}} dt = \int_{r_0}^0 \left(\frac{dt}{dr}\right) dr = \frac{-1}{[2GM]^{\frac{1}{2}}} \int_{r_0}^0 \left[\frac{1}{r} - \frac{1}{r_0}\right]^{-\frac{1}{2}} dr. \tag{4.5}$$

If we define $x \equiv r/r_0$, then

$$t_{\text{ff}} = \left[\frac{r_0^3}{2GM}\right]^{\frac{1}{2}} \int_0^1 \left[\frac{x}{1-x}\right]^{\frac{1}{2}} dx. \tag{4.6}$$

The integral evaluates to $\frac{\pi}{2}$ by substituting $x = \sin^2\theta$, yielding

$$t_{\text{ff}} = \left[\frac{3\pi}{32G\rho_0}\right]^{\frac{1}{2}}, \tag{4.7}$$

where

$$\rho_0 = \frac{M}{(4/3)\pi r_0^3} \tag{4.8}$$

is the initial uniform density. We can apply this equation to our Sun, using the average density $\langle\rho\rangle = 1.4\,\text{g cm}^{-3}$ to get

$$t_{\text{ff}} \approx 0.5\,\text{hours}. \tag{4.9}$$

This is an exceedingly short time! Without internal pressure support, the Sun would collapse in half an hour. Given the immensely longer time over which even human records show the Sun to be maintaining its size, we conclude that the Sun is very effectively in a state of force balance. There must exist a significant source of internal pressure to support the Sun, and other stars, against their self-gravity.

4.2.3 Virial Theorem

If an object is in equilibrium, then we can integrate the force balance Equation (4.3) to get a relationship between total energies. If we multiply Equation (4.3) by $4\pi r^3$ and integrate from 0 to R, we can derive that

$$E_{\text{therm}} = -\frac{1}{2}E_{\text{grav}},\tag{4.10}$$

where

$$E_{\text{therm}} = \frac{3}{2}\langle P\rangle\frac{4}{3}\pi R^3\tag{4.11}$$

is the thermal (or internal) energy of a star, in which angle brackets denote an average value, and

$$E_{\text{grav}} = -\int_0^R \frac{Gm(r)\rho(r)4\pi r^2}{r}dr = -\int_0^M \frac{Gm(r)}{r}dm\tag{4.12}$$

is the gravitational potential energy. For the case of a uniform density sphere, evaluation of the above integral yields

$$E_{\text{grav}} = -q\frac{GM^2}{R},\tag{4.13}$$

with $q = 0.6$. However, stars are centrally concentrated and stellar models yield $q \approx 1.5$ for most stars. Note that q increases as the central concentration increases, but remains of order unity magnitude for all conceivable density distributions.

4.2.4 Total Energy

The total energy of the system can be determined using the virial theorem, yielding

$$E_{\text{tot}} = E_{\text{therm}} + E_{\text{grav}} = \frac{1}{2}E_{\text{grav}}.\tag{4.14}$$

It follows that $E_{\text{tot}} < 0$, which implies that a star is bound, i.e., an amount of energy $|E_{\text{tot}}|$ must be added to a star to break it up and scatter its parts to a large distance.

Now, we can enquire what happens if a star contracts slowly, maintaining near-equilibrium with $E_{\text{tot}} = \frac{1}{2}E_{\text{grav}}$ at all times? If contraction occurs with fixed M and decreasing R, then for some change $\Delta R < 0$, the change in gravitational energy $\Delta E_{\text{grav}} < 0$, while the change in internal energy $\Delta E_{\text{therm}} > 0$. The change in total energy $\Delta E_{\text{tot}} < 0$. Clearly, some of the reduction in E_{grav} is offset by an increase in E_{therm} that results from work done by gravity to compress the star. But achieving a new equilibrium state requires the loss of some energy. Where does it go? The answer is that the energy loss comes from radiation leaving the star. So, a star can radiate (shine) and lose energy, facilitating a slow contraction.

4.2.5 Energy from Gravitational Contraction

In the 19th century, it was popularly believed that stellar radiation and slow gravitational contraction were tied together in the manner described above. Let us pursue this idea to its logical limit. If the Sun contracted from a very large initial state to its current radius R, then

$$|\Delta E_{\text{tot}}| = \frac{1}{2}|\Delta E_{\text{grav}}| = \frac{q}{2}GM^2\left(\frac{1}{R} - \frac{1}{R_{\text{init}}}\right) \approx \frac{q}{2}\frac{GM^2}{R}, \qquad (4.15)$$

since we can safely assume $R \ll R_{\text{init}}$. Given this total available energy, if we divide by the energy loss rate, then we can get an estimate of the lifetime. Assuming that the luminosity L has remained relatively constant, we estimate a lifetime

$$t \approx \frac{|\Delta E_{\text{tot}}|}{L_{\odot}} \approx 0.75\frac{(GM_{\odot}^2/R_{\odot})}{L_{\odot}} \approx 2.4 \times 10^7 \, \text{yr}. \qquad (4.16)$$

Here, we have used values for the Sun: mass $M_{\odot} = 1.99 \times 10^{33}$ g and luminosity $L_{\odot} = 3.83 \times 10^{33}$ erg s^{-1}. This time scale is known as the **Kelvin–Helmholtz time**. This result is fascinating since the Earth's age (estimated to be $\sim 4.5 \times 10^9$ yr from measuring the abundance of radioactive elements in terrestrial, lunar, and meteorite samples) significantly exceeds this value. It tells us that there must be another source of energy in the Sun. Stars are *not* powered by gravity. What else could it be? The answer would emerge in the early twentieth century.

4.2.6 Nuclear Energy

In the 1930s, work by physicists Ernest Rutherford (1871 - 1937), Enrico Fermi (1901 - 1954), and many others, culminating with the work of Frédéric Joliot-Curie (1900 - 1958), established that atomic nuclei could be split with a release of tremendous amounts of energy in accordance with Einstein's theory of special relativity that predicted a mass–energy equivalence. This fission of nuclei is a pathway to energy release from large nuclei because the smaller nuclei have a stronger binding due to the strong nuclear force. Similarly, small nuclei can release energy through fusion to create larger nuclei. The details of the strong nuclear force that is the ultimate source of the binding of nucleons were worked out in succeeding decades, culminating in the successful theory of quantum chromodynamics developed in the 1970s.

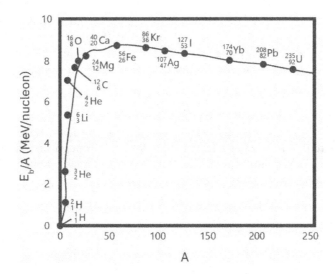

Figure 4.1 The binding energy per nucleon E_b/A of atomic nuclei versus atomic mass A. It increases with increasing A at low values, meaning the nuclei are more tightly bound and more energy (equal to the binding energy) is needed to break the nuclei into its constituent parts (protons and neutrons). The peak value occurs in the "iron group" elements, most notably $^{56}_{26}$Fe. Larger nuclei are less tightly bound. On the left side of the peak, nuclear fusion releases energy, while on the right side of the peak, nuclear fission releases energy. 1 MeV $= 1.602 \times 10^{-6}$ erg.

The binding of protons and neutrons due to the attractive strong nuclear force results in an atom having a negative potential energy. An important measure of the strength of binding is the binding energy (E_b) per nucleon. We can think of ΔE_b as the energy released or absorbed when nucleons (baryons) transform from one nuclear configuration to another. When dropping to a lower energy state, the change in nuclear potential energy E_{nuc} is negative, and ΔE_b ($= -\Delta E_{nuc}$) is positive. Energy is released as a result of the transformation. For nuclei with small atomic mass A, increasing the number of nucleons increases the binding energy per nucleon, as each nucleon can bind with a larger number of neighboring nucleons through the short-range strong nuclear force. The interaction length of the strong nuclear force is only 1 fm $= 10^{-13}$ cm. As nuclei get larger, the strong nuclear force attraction yields diminishing returns, as each nucleon is limited in the number of neighbors that can be packed close to it. Furthermore, the Coulomb repulsion between the protons in the nucleus eventually dominates, as the Coulomb interaction is a long-range force, i.e., decreases with distance r at a much lesser rate (r^{-2}) than the strong nuclear force. See Figure 4.1.

It turns out that fusing elements together gains energy only up to the formation of iron (^{56}Fe). A ^{4}He atom is more tightly bound than four ^{1}H atoms, therefore has a lower (higher) nuclear potential (binding) energy. Therefore, the ^{4}He atom also has a lower mass than the sum of four ^{1}H atoms. As per the theory of special relativity, we know that changes in mass m and energy E are related by

$$\Delta m = \frac{\Delta E}{c^2}, \tag{4.17}$$

where $\Delta E = \Delta E_{\text{nuc}} = -\Delta E_b$ and $\Delta m < 0$ is the mass deficit, equal in magnitude to 0.007 amu per nucleon when four ^{1}H atoms transform to ^{4}He. If the inner 10% of the Sun's mass (the core) is hot and dense enough to undergo fusion reactions, then the total energy available is

$$|\Delta E| = |\Delta m| c^2 = (0.007 \times 0.1 M_\odot) c^2 \approx 10^{51} \text{ erg}. \tag{4.18}$$

If this is the case, then the lifetime of the Sun is

$$t = \frac{|\Delta E|}{L_\odot} \approx \frac{10^{51} \text{ erg}}{3.83 \times 10^{33} \text{ erg s}^{-1}} \approx 10^{10} \text{ yr}. \tag{4.19}$$

The above result finally agrees with the independent estimates of the age of the Earth and solar system. Nuclear energy powers the Sun.

4.2.7 Internal Pressure and Temperature

Equations (4.10) and (4.11) can be combined to yield

$$\langle P \rangle = -\frac{E_{\text{grav}}}{4\pi R^3} = \frac{q}{4\pi} \frac{GM^2}{R^4}. \tag{4.20}$$

Since stellar plasma usually obeys the ideal gas law, we can also estimate

$$\langle P \rangle = \frac{\langle \rho \rangle k \langle T \rangle}{\bar{m}}, \tag{4.21}$$

again utilizing angle brackets to denote average values. We can equate the above two relations and plug in numbers for the Sun, including $\langle \rho \rangle = 1.4 \text{ g cm}^{-3}$ and $\bar{m} = 0.6 \, m_H$ for ionized hydrogen including 10% helium content, to find that $\langle T \rangle \approx 6 \times 10^6$ K. Since the central temperature T_c is expected to be greater than the average temperature $\langle T \rangle$, we conclude that T_c should be high enough ($\sim 10^7$ K) for nuclear reactions to take place. This nuclear energy source will, in turn, maintain the central temperature and pressure required to support the Sun for $\sim 10^{10}$ yr.

4.3 OBSERVING THE STARS

How can we actually measure the distances to stars? How can we estimate their luminosity and temperature? One of the beautiful features of stars is that we can understand them as spherical blackbody radiators in which cooler atoms, ions, or molecules near the surface absorb out certain transition frequencies. We explore these features in the following subsections.

4.3.1 Trigonometric Parallax

In a span of a year, the Earth undergoes a cyclic change in its position in space. Consequently, the observed direction of a star with respect to distant background stars also changes. This shift in the direction is called **trigonometric parallax**. It is the only direct way of finding the distance between a star and the Earth.

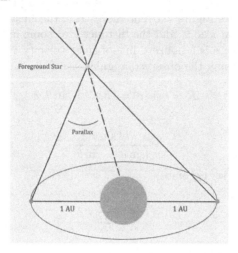

Figure 4.2 An illustration of the principle of parallax. Measurements taken six months apart will show a shift in the angular position of a nearby star relative to the background of much more distant stars.

As seen in Figure 4.2, the parallax angle (in the small angle approximation) is

$$p = \frac{a}{d}, \tag{4.22}$$

where the baseline is $a = 1$ AU, and d is the distance to the star. When the parallax angle p equals one arcsecond ($''$), the distance $d = 1$ pc. In fact, Equation (4.22) provides a definition of 1 pc as the distance at which a lateral distance of 1 AU subtends an angle of $1''$. When we measure p for the nearest star α Centauri, we get $0.76''$, which implies that it is at a distance 1.3 pc. Until late in the 20th century, only about 10,000 stars had their parallaxes measured, with about 500 measured very accurately. The *Hipparcos* satellite, launched in 1989 by the European Space Agency (ESA), was the first space telescope dedicated to astrometry, the precise measurement of the positions, parallaxes and proper motions of stars. Its complete catalog released in 1997 provided parallaxes for 120,000 stars with milliarcsecond accuracy. It also measured another 2.5 million star parallaxes with a lesser degree of accuracy. These gave accurate distances to stars out to several hundred light-years from Earth. The follow-up space satellite *Gaia*, launched by ESA in 2013, is expected to achieve accurate parallax for up to one billion stars, or about 1% of all stars in our Galaxy.

4.3.2 Moving Cluster Method

In the **moving cluster method** of determining distance, we measure the change in the apparent angular size of a cluster over a period of time, which could be decades. See Figure 4.3. The radius R of the cluster can be related to the observed angular size θ' and the distance r at some initial time $t = 0$. At a later time t, if there is a nonzero line-of-sight velocity v, then we can write a similar relation using the observed angular size θ at time t. Therefore

$$R = r \sin \theta' = (r + vt) \sin \theta. \tag{4.23}$$

This leads to

$$r = \frac{[vt \sin \theta]}{[\sin \theta' - \sin \theta]}, \tag{4.24}$$

which for small angles becomes

$$r \cong vt \frac{\theta}{\Delta \theta}. \tag{4.25}$$

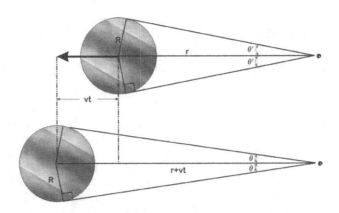

Figure 4.3 An illustration of the moving cluster method.

4.3.3 Luminosity Distances

The flux f is defined as the power (or energy per unit time) that passes per unit area through a surface. Measured at Earth, the flux coming from an astronomical source that is at a distance d and is isotropically radiating with a luminosity L is

$$f = \frac{L}{4\pi d^2}. \tag{4.26}$$

If we know the intrinsic luminosity L, then the measurement of the flux f allows a determination of the distance d. This general principle is utilized repeatedly in astronomy, whenever we find a class of objects for which we think we can determine the intrinsic luminosity L through its correlation with another observable.

4.3.4 Magnitude Scale

The modern magnitude scale for stars is based on a 2000-year-old scale formulated by Hipparchus (190 - 120 BC). In the original scale, the brightest star was given the lowest digit, 1, and the faintest the highest digit, 6. One can think of the brightest star as a first class star, so to speak. In this scale, a five magnitude difference signifies a factor of hundred in brightness. This is a logarithmic scale similar to the human eye, where equal ratios of actual intensity correspond to equal intervals in magnitude.

Keeping the historical scale intact but extending to brighter ($m < 1$) and fainter ($m > 6$) objects, the modern scale is defined by

$$m_2 - m_1 = 2.5 \log_{10} \left(\frac{f_1}{f_2} \right) , \qquad (4.27)$$

where m_2 and m_1 are apparent magnitudes of stars with measured fluxes f_2 and f_1, respectively. The key elements of the magnitude scale are that the scale is logarithmic in nature, brighter objects have smaller magnitudes (even negative values), and the object has relatively low apparent magnitude if it has high luminosity and/or is close to us.

Table 4.1 shows the apparent magnitudes of many prominent objects in the sky, as well as the limiting magnitudes achieved by different visual instruments. The limiting magnitude depends on many factors including the exposure time (only 0.1 s for the human eye), the size of the aperture, and the quantum efficiency (QE), which is the fraction of incident photons that are actually picked up by the detector. The QE ~ 0.01 for the human eye, but ~ 1 for the charge-coupled-devices (CCDs) that are attached to modern telescopes. The limiting magnitude for the *Keck Observatory* listed below is based on a one hour exposure, and the value for the *Hubble Space Telescope* is based on a 22 day exposure. Space-based telescopes do not necessarily have their exposure time limited by the day–night cycle that affects ground-based telescopes.

Table 4.2 illustrates the different ratios f_1/f_2 for objects that differ by fixed increments of apparent magnitude m, illustrating the logarithmic nature of the magnitude scale. The flux ratio becomes unimaginably large as we get to the faintest objects that are detected by major telescopes.

TABLE 4.1 Apparent magnitudes

Object	Magnitude
Sun	−26.5
Full Moon	−12.5
Venus	−4
Jupiter	−2
Mars	−2
Sirius	−1.5
Aldebaran	1
Altair	1
Naked-eye limit	6.5
Binocular limit	10
15 cm telescope	13
10 m Keck telescope	26
Hubble Space Telescope	31

TABLE 4.2 Flux ratios

Δm	f_1/f_2
0.5	1.6
1	2.4
1.5	4
2	6.3
4	49
5	100
6	251
10	10,000
20	100,000,000
25	10,000,000,000

4.3.5 Absolute Magnitude

We know that apparent magnitude depends on distance but we require a true measure of an object's luminosity, i.e., its intrinsic brightness. This quantity is known as **absolute magnitude**. It is defined to be the apparent magnitude of the object if it was located at a distance 10 pc from Earth. If the measured flux from the object at its actual distance d is f_m, and the flux if it were located at a distance of 10 pc is f_M, then using Equation (4.26), we find

$$\frac{f_M}{f_m} = \left(\frac{d}{10\,\mathrm{pc}}\right)^2 . \tag{4.28}$$

This leads us to a definition of the **distance modulus**, which is the difference between the apparent and absolute magnitudes:

$$m - M = 2.5 \log \left(\frac{d}{10 \, \mathrm{pc}} \right)^2 = 5 \log \left(\frac{d}{10 \, \mathrm{pc}} \right). \tag{4.29}$$

In other words, we can determine the absolute magnitude M (i.e., luminosity L) if we measure the apparent magnitude m (i.e., flux f) and know the distance d. Alternatively, for distant objects where parallax (or moving cluster method) is not possible, a physical or empirical model to estimate L (and therefore M) can allow us to estimate the distance d.

4.3.6 Magnitudes at Different Wavelengths

In observational astronomy, photodetectors are usually sensitive to particular wavelength bands. In order to make observations practically viable, we use a standardized system of bands known as the UBV system, consisting of an ultraviolet (U) band centered at 365 nm, a blue (B) band centered at 440 nm, and a visual (V) band centered at 550 nm. In the UBV system, if we assume that a star has a blackbody spectrum, then different wavelengths correlate with different temperatures. If we know the apparent magnitude and distance d, we can determine the absolute magnitude of the stars in each band. Color indices are defined as the difference between magnitudes in two different bands. If the apparent magnitudes in the three bands are labeled U, B, and V, respectively, and the absolute magnitudes as M_U, M_B, and M_V, respectively, then the color indices are

$$U - B = M_U - M_B = 2.5 \log \left(\frac{f_B}{f_U} \right) + \text{constant} \tag{4.30}$$

and

$$B - V = M_B - M_V = 2.5 \log \left(\frac{f_V}{f_B} \right) + \text{constant}. \tag{4.31}$$

The constants can be different in the two equations and are chosen separately so that $U - B = B - V = 0$ for $T = 10^4$ K. Stars with negative color index have temperature $T_e > 10^4$ K and are called **hot stars**, while stars with a positive color index have $T_e < 10^4$ K and are known as **cool stars**.

4.3.7 Bolometric Magnitudes

The **bolometric magnitude** of a star is calculated from the sum of all of its electromagnetic radiation. The total bolometric flux is

$$f_{\mathrm{bol}} = \int f(\lambda) d\lambda. \tag{4.32}$$

Therefore, we can express the apparent bolometric magnitude as

$$m_{\mathrm{bol}} = m_v + 2.5 \log \left(\frac{f_V}{f_{\mathrm{bol}}} \right). \tag{4.33}$$

In usual practice, we measure m_v or M_v for a star and use theoretical models for the inferred type of star to estimate the total luminosity L_{bol} or flux f_{bol}. We can then estimate the bolometric correction (BC) and determine M_{bol} as follows:

$$M_{bol} = M_v + BC, \qquad (4.34)$$

where

$$BC = M_{bol} - M_v = m_{bol} - m_v = -2.5 \log \left(\frac{f_{bol}}{f_V} \right) . \qquad (4.35)$$

It should be noted that the BC is hard to estimate when only a small fraction of the star's energy is radiated in the V band. The blackbody curve peaks in V when $T = 6700$ K.

4.3.8 Stellar Spectra

A star's radiative output can be understood as a blackbody spectrum emerging from the stellar **photosphere**, the layer where photons make the transition from an opaque interior to a transparent exterior. The photosphere is not a perfectly thin layer, and in the course of making this transition, some photons of the continuous blackbody spectrum also get absorbed by cooler atoms, ions, or molecules in the upper layers. Hence, the stellar spectra follow the envelope of a blackbody spectrum of the photospheric effective temperature T_e, but are overlaid with prominent absorption lines. See Figure 4.4. The absorption lines enable us to estimate the chemical abundance at the stellar photosphere if we have a good theory for the energy-level populations and ionization states of the various elements and compounds.

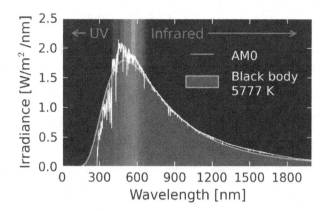

Figure 4.4 The spectrum of the Sun. Solar irradiance as measured above the Earth's mass, is often called AM0, or "atmosphere zero" of attenuation by the Earth. Also plotted is a blackbody spectrum at the effective temperature of the Sun, 5777 K, using a solid angle $2.16 \times 10^{-5}\pi$ steradian for the source (the solar disk). Figure from wikimedia commons, marked as public domain.

Amazingly, the spectrum of a star can allow us to determine almost all information about the physical properties of a star, once we utilize the theory of stellar structure and evolution (see Chapter 5). Stars are classified into several broad categories based on their spectral line patterns. The spectral lines yield information about temperature, luminosity, chemical abundances, velocity, rotation, mass inflow/outflow, magnetic fields, etc. The early classification of stars was done by Annie Jump Cannon (1863 - 1941) at Harvard in the 1900s based on spectra of $\sim 400,000$ brightest stars. In the first try, the order was according to Balmer line strengths, A to P, with A having the strongest Balmer lines and P the weakest. Later, some letters were dropped and the remaining letters were reordered to correspond to a decreasing temperature sequence. See Figure 4.5.

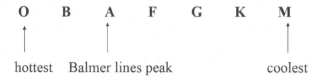

Figure 4.5 The sequence of stellar spectral types, from hottest to coolest. Often memorized by astronomy students using the mnemonic *Oh be a fine girl/guy, kiss me!*

In this classification, each spectral type has 10 subclasses, 0–9, e.g., O0–O9, B0–B9, A0–A9, etc., in which the Sun comes in as a G2 type star. Why do the different spectral classes exist? Recalling the Boltzmann and Saha equations, we find that the $n = 2$ state of hydrogen is most populated at $T \simeq 10^4$ K. Therefore, the strength of the hydrogen Balmer absorption lines, as measured by their equivalent width, is greatest for stars with this photospheric temperature.

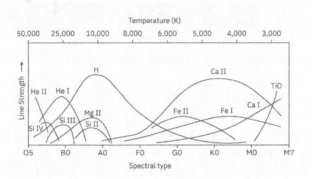

Figure 4.6 The variation of line strengths versus temperature for various neutral and ionic species. Stars of different spectral types therefore show a prominence of lines from different species. The H lines are calculated for the Balmer transitions and other species lines are for their commonly observed transitions. These dependences, with each species peaking at different temperatures, are a result of solving the Boltzmann and Saha equations.

Figure 4.6 shows how Saha's insightful physical theory (Saha, 1921) reveals the explanation for the significant difference in line strengths of different species across the different spectral types of stars. We can convert this information into the realization that the spectral sequences A to M for stars and L, T, and Y for brown dwarfs correspond to sequences of decreasing effective temperature. Furthermore, we can use the theory to determine the relative abundances of the different elements. Without the Saha equation, one may naively think that a G star like the Sun has an abundance of calcium that even exceeds that of hydrogen, given the stronger lines in Ca II. However, the Saha equation shows that the line strength of Ca II is greater than that of the hydrogen Balmer series for G, K, and M stars due to a favorable temperature, even though the abundance of calcium is much less than that of hydrogen.

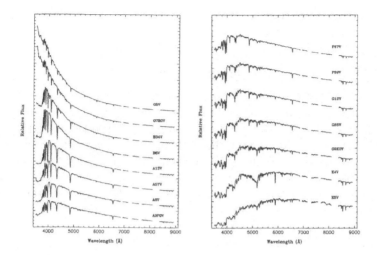

Figure 4.7 A sample of stellar spectra of main-sequence stars (the first letter next to each curve denotes the spectral class). The vertical scale of "Relative Flux" exists only to separate the spectra for clearer viewing. The wavelengths are in Angstroms ($1\text{Å} = 0.1\,\text{nm} = 10^{-8}$ cm). The gaps in the near-IR are excised atmospheric absorption bands. Credit: Silva & Cornell (1992), reproduced with permission from the AAS.

Figure 4.7 shows representative spectra from types O to K. From type O to type A, there is a steady increase in the strength of the hydrogen Balmer lines, with notable transitions Hα (656 nm), Hβ (486 nm) and Hγ (434 nm). For spectral type F and cooler, the strength of the hydrogen lines decreases as collisional excitation of the $n = 2$ state is less likely. Metallic lines now become much stronger, with notable lines being Ca II H and K lines (390 nm), Mg I (515 nm, see especially K stars), and Sodium D (580 nm). The Ca II also produces a triplet of lines at the wavelengths 850 nm, 854 nm, and 866 nm. Listed below are some of the main features of each spectral type in the *Harvard Spectral Sequence*.

O Type Stars: Hottest bluish-white stars; relatively few lines; He II dominates.

B Type Stars: Hot bluish-white stars; more lines; He I dominates.

A Type Stars: White stars; ionized metal lines; hydrogen Balmer lines dominate.

F Type Stars: White stars; hydrogen lines declining; neutral metal lines increasing.

G Type Stars: Yellowish stars; many metal lines; Ca II lines dominate.

K Type Stars: Reddish stars; molecular bands appear; neutral metal lines dominate.

M Type Stars: Coolest reddish stars; neutral metal lines strong; molecular bands dominate.

4.4   HERTZSPRUNG-RUSSELL (H-R) DIAGRAM

Ejnar Hertzsprung (1873 - 1967) and Henry Norris Russell (1877 - 1957) studied the relation between the absolute magnitudes and spectral types of stars. The representation of these quantities is famously called the **H-R Diagram**, which holds a central position in the study of stars. It was initially formulated for nearby stars for which the absolute magnitude (due to parallax) and spectral type were known. About 90% of nearby stars fall on a band known as the **main sequence**, but notable exceptions exist for some spectral types. In these cases, stars can have similar spectra to their counterparts on the main sequence, but have very different luminosities.

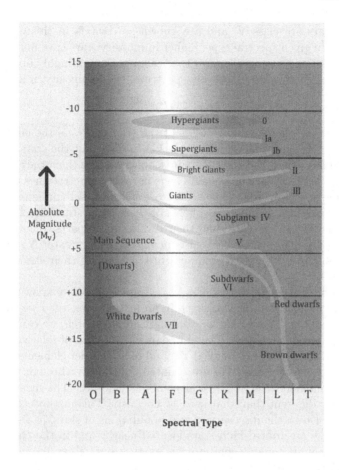

Figure 4.8 A schematic representation of an H-R diagram that highlights the different bands where stars are found, according to the Morgan–Keenan (M–K) classification scheme. The main sequence stars are classified by the Roman numeral V. The M spectral class (and some of the L class) covers the main sequence red dwarfs, which do not have a separate M-K classification. The L and T spectral classes contain the brown dwarfs, which are cooler than main-sequence stars and also do not have a separate M-K classification.

4.4.1 Luminosity Classes

The empirical Morgan–Keenan (M–K) classification scheme is used to group stars into luminosity classes that are shown in Figure 4.8. These descend in Roman numerals from I to VII from the most luminous class to the faintest class, which are actually stellar remnants known as **white dwarfs**. The main sequence stars are class V, and are known as **dwarfs** in this classification scheme. For a given spectral type, higher luminosity stars have much narrower spectral lines, and stronger ionized species lines. Since a highly opaque object such as a star radiates a blackbody spectrum, the luminosity of a star is

$$L = 4\pi R^2 \sigma T_e^4 , \qquad (4.36)$$

where R is the (photospheric) radius of the star and T_e is the effective temperature of the photosphere. This equation implies that the stars in classes I to IV are larger than the class V main sequence stars; thus, they are labeled as **giants**. Furthermore, these giants have narrower spectral lines than dwarfs of the same spectral type. This is due to less collisional (pressure) broadening of spectral lines in the atmospheres of these giant stars that have weaker surface gravity and pressure than their dwarf counterparts. A lower electron density (N_e) in their atmospheres also means there is a greater N_+/N_0 ratio in the Saha equation for a given temperature, explaining their stronger ionized species lines.

What about the subdwarfs (class VI)? Empirically, if we know the temperature and gravitational field strength in an atmosphere, then we can predict the line strengths. Then, if the line strengths don't match, we can change the abundance until they match. The abundance is the sum of relative mass fractions of H (called X), of He (called Y), and of all heavier elements (called Z). The elements heavier than He are counted collectively through the number Z and astronomers refer to these elements as "metals." The meaning of this word is very different than that used by chemists! This method of estimating abundances led to the discovery of two populations of stars, as listed below.

Pop I: These are **metal-rich** stars located nearby and in the Galactic disk. They have similar metal composition as the Sun, $Z \approx 0.02$, along with $X \approx 0.73$ and $Y \approx 0.25$.

Pop II: These are **metal-poor** stars and are older and located in the Galactic halo. They have $Z \approx 0.001$. These stars would have formed early in the history of the Galaxy and have $X \approx 0.75$ and $Y \approx 0.25$.

Subdwarfs represent a main sequence for Pop II stars. The presence of fewer metals makes them appear slightly hotter and bluer than their Pop I counterparts. Metals provide many channels for absorbing photons, so less metals means a lower **opacity** (this term is defined in Chapter 5). Less absorption of photons reduces the **radiation pressure** inside the star, causing the star to be smaller and hotter than its Pop I counterpart. The overall effect on luminosity from Equation (4.36) is that it is decreased from that of its Pop I counterpart. These effects move the Pop II main sequence to a position a little to the left of the Pop I main sequence.

We note that there are also the hypothetical Pop III stars that are thought to be the first stars that formed after the Big Bang, and would have no metal content at all. Although some stars have been found in our Galaxy with exceedingly low metal content, no stars with spectra that are completely metal-free have yet been detected.

4.4.2 Color-magnitude diagrams

Color-magnitude diagrams provide a quicker route to making an H-R diagram. Since color indices (e.g., $B - V$) correlate roughly with spectral type or temperature, as described in Section 4.3.6, we can plot color index versus absolute magnitude as a proxy for temperature versus luminosity. See Figure 4.9. Although colors are much easier to measure than temperature or spectral type, this method still has pitfalls, since $B - V$ is not a good indicator for very hot or very cool stars.

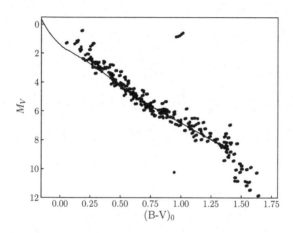

Figure 4.9 Color-magnitude diagram for the Hyades open cluster. Adapted from data presented by Johnson et al. (1962).

4.5 STAR CLUSTERS

Star formation in large gas clouds held together by self gravity usually yields star clusters. Empirically, there are two types of stellar clusters in our Galaxy.
Open clusters: These clusters have about $\sim 10^2$–10^4 stars and are located close to the Galactic plane and consist of Pop I stars. See Figure 4.10.
Globular clusters: These clusters have $\sim 10^5$–10^6 stars and can be found at large heights above the Galactic plane. They consist of Pop II stars and are the oldest observed objects in the universe. See Figure 4.11.

Figure 4.10 The open cluster Trumpler 14. It is one of the largest open clusters in the Galaxy, and located at a distance of about 2.8 kpc. About 2000 stars reside within the cluster, and range in mass from about 0.1 $M_\odot$ all the way up to several times 10 $M_\odot$. The cluster contains one of the highest concentrations of massive, very luminous stars in our Galaxy. A small dense concentration of gas and dust is also seen as a dark spot in silhouette, a little to the left and below the center of the image. Credit: NASA, ESA, and J. Maíz Apellániz.

There are about 1100 known open clusters in the Galaxy, with the Pleiades and Hyades clusters being the two closest and most prominent. Open clusters are located near the plane of the Galaxy and provide evidence for ongoing star formation. The globular clusters span a spherical volume with radius ~ 15 kpc around the Galactic center, and about 150 have been detected. The globular clusters provided the first evidence that the Sun was not at the center of our Galaxy or of the universe for that matter. They were formed when the Galaxy was very young, and have ages from 8 to 12 Gyr. Fundamental reasons for these two distinct types of star cluster in our Galaxy are yet to be established.

Figure 4.11 The globular cluster M 13. Containing several hundred thousand stars and located at a distance of 6.8 kpc from Earth, M 13 is seen in the constellation Hercules and is one of the oldest objects in our Galaxy, with an estimated age 11.65 Gyr. In 1974, the radio telescope at the *Arecibo Observatory* in Puerto Rico was used to beam a focused radio message toward M 13, in one of the most notable examples of the *Search for Extraterrestrial Intelligence* (SETI) project. The message will reach the hypothetical alien civilizations in this densely packed environment of stars in about 20,000 years. Credit: Adam Block, www.AdamBlockPhotos.com.

4.5.1 Distance Determinations to Clusters

The spectral type can be used to estimate the absolute magnitude M of a star. For example, if a star is determined to be a main-sequence G star, we can use a calibrated H-R diagram to determine the corresponding absolute magnitude M. The distance modulus formula, Equation (4.29), can then be used to calculate the distance to the star. This method is known as **spectroscopic parallax**. It is best done with data for an entire star cluster, and in this case is known as **main-sequence fitting**. If we find the offset between m and M for an entire cluster main sequence, we can get the distance d to the cluster, again using Equation (4.29). The large number of data points minimizes the errors. Ultimately, the main-sequence fitting relies on a well-calibrated H-R diagram, e.g., from parallax or the moving cluster method.

CHAPTER 5

Stellar Evolution

5.1 INTRODUCTION

The development of a theory of stellar evolution was one of the great successes of 20th century astrophysics. Based on a surprisingly simple system of spherically symmetric equations of mechanical equilibrium, energy transfer, and energy generation, it can account for many of the observed properties of stars and their positions on the H-R diagram. The physical effects in these models include gravity, thermal pressure, radiation pressure, nuclear reactions, convection, and the quantum mechanical effects of zero-point motion. Most of stellar evolution can be understood using a **quasistatic** model, i.e., one in which a star evolves from one state of mechanical equilibrium to another. The constraints of thermodynamics mean that stars are always radiating energy away to their environment, and this can be understood as a primary driver of their evolution. The nuclear reactions that occur in the cores of stars also drive their evolution by maintaining an energy balance even as they gradually change the composition of the star.

It turns out that the evolutionary history of a star depends on only two parameters: its initial mass M and its initial composition μ. Other initial conditions like the rotation rate or magnetic field play only a peripheral role. The vast majority of a star's life is spent on the main sequence (MS), when the star fuses hydrogen into helium in its core. Indeed, some 90% of all observed stars are found to lie on the MS.

5.2 STELLAR STRUCTURE AND EVOLUTION

We know from an estimate of the free-fall time (≈ 30 min for the Sun) of stars that their very much longer lifetime implies that they are exceedingly close to a perfect hydrostatic equilibrium. So, why do they evolve?

We first consider the case of a pre-main-sequence (PMS) star, ignoring the effect of nuclear reactions for the moment. The laws of thermodynamics tell us that heat always flows from hotter to cooler objects. A star (whether PMS or MS), surrounded by the coldness of space, is perfectly set up to do

DOI: 10.1201/9781003215943-5

this. The heat flow occurs in the form of blackbody radiation that leaves the stellar surface. The time scale to lose a significant amount of the internal store of thermal energy is $\approx 10^7$ yr, much longer than the free-fall time, so the object can easily readjust to a new state of equilibrium. The virial theorem tells us that for a total energy change $\Delta E_{tot} < 0$, the gravitational energy changes by $\Delta E_{grav} = 2\Delta E_{tot} < 0$, while the thermal energy changes by $\Delta E_{therm} = -\Delta E_{tot} > 0$. Hence, the thermal energy actually increases, while the gravitational energy decreases as the star contracts slightly. A star is a strange object indeed: we say that it has a *negative* specific heat since the ratio $\Delta E_{tot}/\Delta T$ is negative. This model of slow contraction driven by surface energy losses is exactly the model for MS stars that was developed by Kelvin and Helmholtz in the 19th century. But, the implied lifetime of the Sun is then $\approx 10^7$ yr, too short in comparison to estimates from geological records on Earth. We now know that this kind of evolution is only appropriate for PMS stars.

The MS is characterized by a new energy source: nuclear reactions, which occur at a sufficient rate to supply energy to balance the surface energy losses. Calculations show that this occurs when the object has a mass that exceeds the substellar mass limit, $\approx 0.075\, M_\odot$ (Chabrier & Baraffe, 2000). A main sequence star settles down to a steady state that lasts as long as hydrogen is available in the core to fuse into helium.

Modeling of the MS starts with the idea that a star is in an initial hydrostatic equilibrium with uniform composition $\mu(r) = \mu = $ constant. When the central temperature and density allow it to proceed, nuclear energy generation starts at a rate $\epsilon(r)$, thereby changing the composition $\mu(r)$ of the star. Any change in composition changes the opacity $\kappa(r)$ of the system and therefore affects the pressure $P(r)$, temperature gradient dT/dr, etc. These changes lead to the star adjusting to a new equilibrium. This cycle of readjustment continues until the star runs out of energy sources. Figure 5.1 shows how the evolutionary models proceed.

An interesting sidelight on stellar evolution is that, unlike nuclear reactions on Earth, stars that are supported primarily by thermal pressure have a natural built-in "safety valve" for their nuclear reactions. Stars don't blow up due to nuclear reactions! Why? Any increase in the nuclear energy generation rate beyond that required for the radiative and hydrostatic equilibrium will lead to a temperature increase. A gas that is supported by thermal pressure will expand in response, thereby lowering the temperature and reducing the nuclear energy generation rate that is a steep function of temperature. In this way, the star will return to its original equilibrium state. By this and the analogous reasoning if the nuclear energy generation rate is initially reduced, we can say that the nuclear burning in a star supported by thermal pressure is stable.

$\varepsilon(r) \;\;\Rightarrow\;\;$ changes composition $\mu(r)$

$\Rightarrow\;\;$ changes opacity $\kappa(r) \;\;\Rightarrow\;\;$ affects $P(r), \dfrac{dT}{dr}$, etc.

$\Rightarrow\;\;$ star adjusts to new equilibrium

$\Rightarrow\;\;$ affects $\varepsilon(r)$, so back to first step....

Figure 5.1   A flowchart for the calculation of quasistatic stellar evolution models.

In stellar structure equations, the outer radius R of the star is where the density drops to zero. In reality, the star has a tenuous outer atmosphere, and the radius is associated with the photosphere, where the photons emerging from the interior make the transition from a random-walk transport to free-streaming at the speed of light. This is usually estimated as the radial location where an inward integration from the exterior yields an optical depth $\tau = 2/3$. The study of **stellar atmospheres**, including the detailed radiative transfer calculations of the emergent spectral line shapes and intensities, is a subject in itself and we do not cover it here.

5.2.1   Hydrostatic Equilibrium

Consider an interior spherical shell of radius r and thickness dr. In a situation of hydrostatic equilibrium, a pressure differential dP across this shell exerts a net force that balances the force of gravity. Since the pressure is applied to the spherical surface of area $4\pi r^2$, we can write

$$4\pi r^2 dP = -\frac{GM(r)\rho(r)4\pi r^2 dr}{r^2}, \tag{5.1}$$

where $\rho(r)$ is the local mass density and $M(r)$ is the enclosed mass at radius r. The above equation can be expressed as

$$\frac{dP}{dr} = -\frac{GM(r)\rho(r)}{r^2}. \tag{5.2}$$

This is the equation of hydrostatic equilibrium, in which the force of gravity per unit volume is counteracted by the pressure gradient.

5.2.2   Mass Continuity

The mass dM contained within a spherical shell of thickness dr located at radius r is $\rho(r)4\pi r^2 dr$, which implies

$$\frac{dM}{dr} = 4\pi r^2 \rho(r). \tag{5.3}$$

This equation defines the mass distribution $M(r)$ that appeared in Equation (5.2).

5.2.3 Equation of State

An equation of state describes the relation between different variables that describe the state of matter under a given set of conditions. For example, the perfect (or ideal) gas law describes the relation between the pressure, temperature, and density of a gas. It is a reasonable approximation for a wide range of pressures and temperatures. Here, we illustrate how to close the system of stellar structure equations by using the perfect gas law,

$$P(r) = n(r)kT(r), \tag{5.4}$$

although detailed modeling must also include **radiation pressure** ($P_{\rm rad} = aT^4/3$) and **degeneracy pressure** ($P_{\rm deg} \propto \rho^{5/3}$; see Chapter 7) . The latter is a pressure due to the zero-point motions associated with quantum mechanics. At sufficiently high densities, the electrons in a plasma reach a state where the degeneracy pressure associated with their zero-point energy can exceed the thermal pressure. Such objects are called **degenerate**. At even higher densities, nucleons can also become degenerate. In Equation (5.4), the number density can be expressed as

$$n(r) = \frac{\rho(r)}{\mu(r)m_H}, \tag{5.5}$$

where the composition is

$$\mu(r) = \frac{1}{2X + \frac{3}{4}Y + \frac{1}{2}Z}, \tag{5.6}$$

and $\mu(r)m_H$ is the mean particle mass. In the above equation, X, Y, and Z are the mass fractions of the gas in hydrogen, helium, or metals, respectively. The numerical factors in front of each come from an assumption that the atoms have been fully ionized and that the metals usually contain twice as many nucleons as electrons. Finally then, we can write

$$P(r) = \frac{\rho(r)kT(r)}{\mu(r)m_H}. \tag{5.7}$$

5.2.4 Energy Conservation

Similar to the mass continuity equation, we can implement the idea of energy conservation by requiring that the luminosity $L(r)$, measured in erg s^{-1}, only changes by an amount dL within a spherical shell of thickness dr if it contains an energy source. If the energy generation rate is $\epsilon(r)$, measured in erg g^{-1} s^{-1}, then

$$\frac{dL}{dr} = 4\pi r^2 \rho(r)\epsilon(r). \tag{5.8}$$

The energy generation rate $\epsilon(r)$ would be zero in a perfectly static star undergoing no nuclear (or chemical) reactions. In a MS star, ϵ comes from hydrogen to helium conversion through the **p-p chain** or the **CNO cycle** that is described later. In the post-main-sequence phase, ϵ is attributed to further nuclear fusion reactions that lead to heavier elements.

At this point in our discussion, equations (5.2), (5.3), (5.7), and (5.8) constitute a system of four equations with six unknowns, M, ρ, P, T, L, and ϵ. The system will be closed and a solution can be found, if in addition to specifying ϵ, one can relate the temperature $T(r)$ to other physical variables. Note that in many cases, we are presenting simplified equations in order to present a relatively compact closed system of equations. Modern calculations will often include more physical processes through more complicated equations or tabulated values.

5.2.5 Energy Transport

An equation for energy transport can relate the temperature gradient dT/dr to other physical variables and is required to close the system of equations of stellar structure. Physically, this means that we need to specify the mechanism of energy transport. The energy transport from hotter to cooler regions can be accomplished in three ways as discussed below.

5.2.5.1 Conductive Transport

Conductive transport occurs due to collisions between particles. The ability to transport heat, the conductivity, is proportional to the product of the mean free path (mean collision distance) and the typical speed of the particles. Since the mean free path of particles is smaller than that of photons along with the mean speed of particles being much less than the speed of light, we can generally ignore conductive transport in comparison to radiative transport by photons. However, in the degenerate cores of evolved stars, the conductive transport due to electrons can become dominant.

5.2.5.2 Radiative Transport

Since the mean free path ℓ of a photon is very small compared to the distance to travel to its surface, we can treat the transport of radiant energy as a diffusive process. Here, we derive an expression for dT/dr in an approximate manner using some simplifying assumptions.

Assuming the principle of local thermodynamic equilibrium (LTE), we can say that each layer of the star has achieved a blackbody spectrum at its local temperature. In that case, the flux of energy F can be related to temperature T using the Planck theory, so that

$$F = \sigma T^4 \Rightarrow dF = 4\sigma T^3 dT . \tag{5.9}$$

Furthermore, we can use the theory of radiative transfer to write that

$$dF = -\kappa(r)\rho(r)F(r)dr\,, \tag{5.10}$$

where κ is the frequency-integrated (gray) opacity.

If we combine the above equations, use $L(r) = 4\pi r^2 F(r)$, and include a correction factor of $3/4$ from a more complete treatment of radiative transfer, we get

$$\frac{dT}{dr} = -\frac{3\kappa(r)\rho(r)}{64\pi\rho r^2 T^3(r)}L(r)\,. \tag{5.11}$$

This is the equation for radiative transport, relating the local temperature gradient dT/dr to the luminosity and other physical variables at a given radius.

The opacity κ of stellar matter is a measure of its ability to scatter or absorb radiation. In practice, it is a very difficult quantity to calculate, as it requires a detailed estimate of the ionization levels, energy-level populations in all atomic and molecular species, quantum mechanical estimates of cross sections, integration over the photon frequency distribution, etc. Scattering is defined as the conversion of an incoming photon into an outgoing photon whose direction is correlated with that of the incoming photon. Absorption is a process where the incident photon disappears entirely (converted to another form of energy) or a new photon appears that is not correlated in direction with the incident photon. To simplify our thinking, we consider the four general sources of opacity below.

Electron scattering: An interaction of a photon with a free electron. The nonrelativistic elastic scattering of a photon by an electron is known as Thomson scattering. It is characterized by the classical Thomson cross section

$$\sigma_{\mathrm{T}} = \frac{8\pi}{3}\left(\frac{e^2}{m_e c^2}\right)^2\,. \tag{5.12}$$

Free-free absorption: The absorption of a photon by a free electron in the presence of a heavy nucleus. The reaction can be written in symbolic form as

$$e^- + {}^A_Z\mathrm{X} + \gamma \rightarrow e^- + {}^A_Z\mathrm{X}\,, \tag{5.13}$$

where γ represents a photon and X is a nucleus with atomic number Z and atomic weight A. An electron cannot absorb a photon and satisfy both momentum and energy conservation, but the presence of the nucleus allows the conservation laws to be satisfied in the three-body interaction. You can think of this as the inverse process of when an electron passes by a positively charged nucleus, is accelerated, and releases a photon.

Bound-free absorption: This is an ionization process. The electron absorbs the photon and moves from a bound state to a free state.

Bound-bound absorption: The electron makes a transition from one bound energy level to another. A collisional de-excitation typically takes place, and if radiative de-excitation takes place, it can involve photons of different frequencies than the incident photon. In the limit of radiative de-excitation to the original level, unaffected by any collisional interaction, we may consider this to be a scattering process.

It is difficult to have an elegant equation to describe opacity that describes all the processes and is integrated over all frequencies. Modern calculations use a table of values that is compiled from detailed modeling of all known interactions. Here, we point out that in 1923 Hendrik Kramers (1894 - 1952) simplified the already existing complex equations and reduced the opacity to a frequency-integrated form that depends on temperature and density. The **Kramers opacity** can be written as

$$\kappa = \text{constant} \times Z(1 + X)\frac{\rho}{T^{3.5}}, \tag{5.14}$$

and is based on considerations of free-free and bound-free absorption. Notably, this formula shows that the temperature T has a much stronger effect on the opacity than does the density ρ. The strong inverse temperature dependence arises because a higher temperature means greater speeds of particles and a lesser interaction time (of close approach) for free-free or bound-free interactions. Notably, the Kramers opacity usually leads to a decreasing opacity when the temperature of a star rises, and in a situation of compression, the temperature rise will dominate the density effect and the opacity will decrease. This means that the rebound toward the equilibrium state will happen with a little less radiation pressure. This naturally damps any oscillations about the equilibrium state. In Section 5.4.5, we will discuss what happens when this situation is changed.

Finally, we note a posteriori that the photon mean free path ℓ within the Sun is indeed incredibly small, thereby justifying the use of the radiative diffusion limit. Electron scattering alone yields a typical average opacity $\kappa \approx 1$ cm^2 g^{-1}, and given that the mean density of the Sun is $\rho \approx 1$ g cm^{-3}, we estimate an average value $\ell = 1/(\kappa\rho) \approx 1$ cm! Inclusion of other sources of opacity reduces this quantity even further.

5.2.5.3 *Convective Transport*

Convection is a means of transport of thermal energy due to the macroscopic motion of convective "eddies." The eddies are macroscopic parcels of gas that can rise when perturbed upward because they have a greater temperature than their surroundings, or descend when perturbed downward because they are cooler than their surroundings. Focusing on the rising eddies, we note that they will exchange heat with their surroundings and cool down, then descend. This ongoing cycle of rising and descending eddies is known as convection. It can set in when the temperature gradient in an atmosphere is high enough to exceed the temperature gradient implied by adiabatic displacements. Quantitatively, this is written as

$$\left.\frac{dT}{dr}\right|_{\text{atmospheric}} > \left.\frac{dT}{dr}\right|_{\text{adiabatic}}. \tag{5.15}$$

In stars, the atmospheric temperature gradient is normally set by radiative transport, as in Equation (5.11). Very high (superadiabatic) temperature gradients can occur due to high opacity, e.g., in the cooler outer parts of low-mass stars, or due to the presence of a very strong heating source in the core, as happens in high-mass stars. These causes of a high value of dT/dr can be seen in Equation (5.11).

The two different pathways to convection are why low-mass and high-mass stars have different locations for their convection zones. See Figure 5.2. Low-mass stars have a convective outer region coupled with a radiative inner region, while more massive stars (above about 1.5 $M_\odot$) have a convective inner region coupled with a radiative outer region. On the other hand, very low-mass stars can be convective throughout their interior, as can pre-main-sequence stars. In the post-main-sequence phase of low-mass stars, the expanded cool outer envelope develops a high opacity and becomes convective.

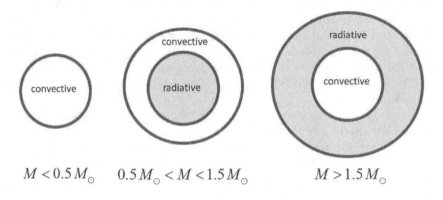

$$M < 0.5\,M_\odot \qquad 0.5\,M_\odot < M < 1.5\,M_\odot \qquad M > 1.5\,M_\odot$$

Figure 5.2 Illustration of the mechanisms of energy transport within stars in different mass ranges. The location of the boundary between radiative and convective zones, when both exist, depends on the mass of the star. For the Sun the boundary is at $r/R_\odot \approx 0.7$.

What happens when convection actually sets in? The nonlinear outcome of convective instability, the rise and fall of all those convective eddies, is that it is very efficient and brings the temperature gradient of the atmosphere back to being close to the marginally stable state of an adiabatic temperature gradient. For adiabatic displacements, thermodynamics tells us that the temperature-pressure relation is

$$T \propto P^{(\gamma-1)/\gamma}, \tag{5.16}$$

where γ is the ratio of specific heats (at constant pressure and at constant volume) of the gas, and is equal to 5/3 for a monatomic gas. The above equation leads to

$$\frac{d\ln T}{d\ln P} = \frac{\gamma-1}{\gamma}. \tag{5.17}$$

Next, using the identity

$$\frac{dT}{dr} \equiv \frac{d\ln T}{d\ln P}\frac{T}{P}\frac{dP}{dr}, \tag{5.18}$$

and inserting equations (5.17) and (5.2), we finally arrive at

$$\frac{dT}{dr} = -\left(\frac{\gamma-1}{\gamma}\right)\frac{\rho T}{P}\frac{GM}{r^2}. \tag{5.19}$$

The above equation replaces Equation (5.11) in the stellar structure equations whenever the condition given by Equation (5.15) is satisfied; in other words when the magnitude of dT/dr given by Equation (5.11) exceeds that given by Equation (5.19).

5.2.6 Energy Sources

Nuclear fusion requires high temperatures so that positively charged nuclei of hydrogen or other elements can get close enough for the short-range **strong nuclear force** to bind them together. The strong nuclear force is attractive on length scales of $\approx 10^{-13}$ cm and dominates the Coulomb repulsion at these scales. Classical estimates using the Maxwell-Boltzmann distribution show that two protons need to have random speeds corresponding to a temperature of $\approx 10^{10}$ K in order to overcome the Coulomb barrier. This seems insurmountable, since even stars cannot achieve such high temperatures in their centers. The solution is provided by quantum mechanics, which shows that quantum tunneling can allow the protons to coalesce even when the temperature is $\approx 10^7$ K! Stars shine due to this mysterious quantum effect. The first step in converting hydrogen to helium is the merger process of two protons (also labeled as ${}^1_1\text{H}$):

$${}^1_1\text{H} + {}^1_1\text{H} \longrightarrow {}^2_1\text{H} + e^+ + \nu_e. \tag{5.20}$$

This first step, the proton–proton merger, creates a deuterium nucleus ${}^2_1\text{H}$ along with a positron (e^+) and an electron neutrino (ν_e). This step gives the process its name, the *p-p* chain. The process involves the **weak nuclear force**, which has much lesser strength and interaction length than the strong nuclear force. The weak force is needed to accomplish the conversion of a proton into a neutron, positron, and electron neutrino. This transformation is known as **inverse beta decay** and is actually endothermic, requiring an energy of at least 1.8 MeV. However, the deuterium nucleus has binding energy 2.225 MeV, which makes the overall reaction exothermic. The involvement of the weak nuclear force makes this first step in the p-p chain rather inefficient. Subsequent steps are mediated by the strong nuclear force and proceed on a shorter time scale. They can occur in three different pathways. Here, we note that the most common subsequent steps in the p-p chain are:

$$^2_1\text{H} + {}^1_1\text{H} \longrightarrow {}^3_2\text{He} + \gamma, \tag{5.21}$$

$$^3_2\text{He} + {}^3_2\text{He} \longrightarrow {}^4_2\text{He} + {}^1_1\text{H} + {}^1_1\text{H}, \tag{5.22}$$

where γ represents a photon. Clearly, the first two parts of this chain must occur twice for each instance of the third part. The overall reaction then utilizes six hydrogen nuclei to produce one helium nucleus plus two free hydrogen nuclei. The net effect can be summarized as

$$^1_1\text{H} + {}^1_1\text{H} + {}^1_1\text{H} + {}^1_1\text{H} \longrightarrow {}^4_2\text{He} + e^+ + \nu_e + \gamma. \tag{5.23}$$

The positron will quickly annihilate when it encounters its antiparticle, an electron, thereby creating another photon. The neutrino will generally leave the star without any interactions. The photons released in the p-p chain are gamma rays due to the relatively large energy gaps between the different nuclear configurations. However, the photons quickly equilibrate to a Planck distribution of energies in each layer of the star, at the local temperature T, due to the many absorptions and re-emissions that take place. The main branch of the p-p chain discussed here releases a total of 26.2 MeV of energy per reaction chain, which includes the effect of the annihilation of the positron but not the energy carried away by the neutrino.

It is useful to have an approximate expression for the energy production rate in the p-p chain that can be used to close the system of stellar structure equations. It turns that the reaction rate is approximately proportional to T^4 in the appropriate temperature range, and since the nuclear reactions are a binary process of collisions of protons of number density n_p, the rate is also proportional to $n_p^2 \propto X^2\rho^2$. We can close the system of equations with an expression of the form

$$\epsilon = \text{constant} \times X^2\rho^2 T^4. \tag{5.24}$$

Hans Bethe (1906 - 2005) realized that since the p-p chain has an inefficient means of getting started, other channels of hydrogen to helium fusion could be relevant. He found that the same net conversion summarized by Equation (5.23) could be accomplished by a series of reactions in which carbon nuclei act as a catalyst, i.e., they participate in the reaction chain but at the end there are no more or less of them than before. Unstable isotopes of nitrogen and oxygen are also produced and destroyed during the cycle. These heavier elements have to already exist as a result of nucleosynthesis in a previous generation of stars whose remnants made their way into the star-forming gas cloud of the present star. This carbon–nitrogen–oxygen (CNO) cycle requires slightly higher temperatures to get started because of the increased Coulomb barrier of a proton encountering a nuclei of carbon (charge $+6e$), nitrogen ($+7e$), or oxygen ($+8e$) during the cycle. However, for stars of mass about $1.5\,M_\odot$ and above, it becomes the dominant energy production mechanism. It is less dependent on the weak interaction and is a much more sensitive function of temperature than the p-p chain. Its prodigious output at high temperatures is the reason that these stars develop convection zones in their cores. We do not deal with the quantitative details of the CNO cycle here.

5.3 MASS–LUMINOSITY RELATION AND LIFETIME

The mass–luminosity relation is determined through observations of binary star systems, as described in Chapter 6. The steep dependence on mass of the luminosity of a MS star can be understood directly from the stellar structure equations. We find scaling relations by replacing derivatives with differences between center and surface, taking some surface quantities to be negligible in comparison to the central value, and using average interior values for some quantities. This is a useful technique in astrophysics that allows for quick estimates. The equation of hydrostatic equilibrium is transformed as follows:

$$\frac{dP}{dr} = -\frac{GM(r)\rho(r)}{r^2} \Rightarrow \frac{0 - P_c}{R} \propto -\frac{M\rho}{R^2} \Rightarrow P_c \propto \frac{M\rho}{R}, \tag{5.25}$$

where P_c is the central pressure, M is the total mass, R is the outer radius, and ρ is the average density. For a perfect gas, we know that

$$P_c \propto \frac{\rho T_c}{\mu} \Rightarrow \rho T_c \propto \frac{\mu M \rho}{R} \Rightarrow T_c \propto \frac{\mu M}{R}, \tag{5.26}$$

where the subscript "c" designates a central value. Using the same set of approximations to the radiative energy transfer equation, we find

$$L \propto \frac{R^4 T_c^4}{\kappa M}, \tag{5.27}$$

therefore

$$L \propto \frac{R^4}{\kappa M} \cdot \left(\frac{\mu M}{R}\right)^4 \propto \frac{\mu^4 M^3}{\kappa}. \tag{5.28}$$

Observed results for the MS yield a similar mass–luminosity relation, with an empirical good fit for the entire mass range being

$$L \propto M^{3.5}, \tag{5.29}$$

although it is a bit shallower at the low mass end and steeper for intermediate mass stars. From the above calculations, we can conclude that massive stars are more luminous but have shorter lifetimes than less massive stars. The lifetime can be estimated as

$$t \propto \frac{M}{L} \propto M^{-2.5}\,\kappa\,, \qquad (5.30)$$

assuming that a roughly fixed fraction of the stellar mass can be converted from hydrogen to helium. Using the empirical Equation (5.29) and our earlier estimate of a MS lifetime $\approx 10^{10}$ yr for a 1 $M_\odot$ star, we estimate the MS lifetime of a star of any mass M to be

$$t \approx 10^{10} \left(\frac{M}{M_\odot} \right)^{-2.5} \text{yr}. \qquad (5.31)$$

By this formula, a high mass star with $M = 25\,M_\odot$ has a MS lifetime of only 3×10^6 yr, while very low mass stars tend to have MS lifetimes longer than the ≈ 14 Gyr age of the universe.

The above lifetimes mean that all M dwarfs that were ever created in the universe are still in their MS phase. In contrast, massive stars must have formed very recently in Galactic history, and therefore reveal the locations of recent star formation.

5.4 EVOLUTIONARY STAGES

The equations (5.2), (5.3), (5.6), (5.7), (5.8), (5.11) or (5.19), (5.14), and (5.24) form a closed set that can be solved for an equilibrium stellar structure and evolved in time due to the effect of nuclear energy generation that gradually changes the composition, which, in turn, affects all the other stellar properties. Astronomers verify the theory by comparing the models of evolution of stars of different mass and composition with the overall demographic snapshot of stars of different ages, masses, and compositions.

5.4.1 A Snapshot of the Stellar Populations

Once the distance to a star is known by parallax or other means, it can be placed on the H-R diagram that relates absolute magnitude to spectral type. In terms of physical variables, the H-R diagram is a plot of luminosity L versus effective temperature T_e, in which T_e increases from right to left on the abcissa. Figure 5.3 shows an H-R diagram with the positions of a large number of prominent stars labeled. The **white dwarf** branch is located well below and to the left of the main sequence; these are the remnants of low mass stars that remain after nuclear fusion ceases. The **instability strip** cuts across the H-R diagram and covers the region of parameter space where the opacity in the envelope is expected to increase with increasing temperature, leading to the development of large-scale nonlinear oscillations. These **variable stars** include the Cepheid variables, which are a Pop I class of supergiant stars that exhibit pulsations of their surface layers due to internal opacity effects, and their Pop II counterparts known as W Virginis stars.

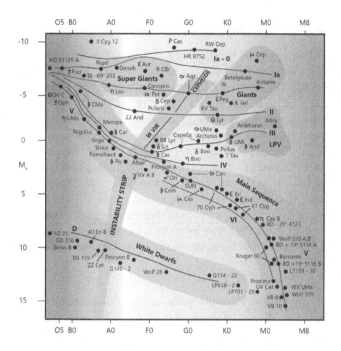

Figure 5.3 The H-R diagram with many famous stars and white dwarfs identified by name and location. The instability strip is a region within which the opacity effects allow for nonlinear pulsations to develop, leading to well-known categories of variable stars such as Cepheids and W Virginis stars.

5.4.2 Evolution of Very Low Mass Stars

The M dwarfs have cool interiors with relatively high opacity and the interiors can be fully convective. At masses above $0.5\,M_\odot$, an inner region is warm enough and the opacity low enough that the temperature gradient never becomes superadiabatic, so that region is governed by radiative energy transport. An outer region is cool and convective.

The MS lifetime of an M dwarf is $\approx 10^{12}$ yr, much longer than the age of the universe, so all M stars created in the 14 Gyr history of the universe are expected to still be on the main sequence. These stars are fully convective, so in addition to having a low nuclear reaction rate due to a low T_c, the convection can bring hydrogen from the outer layers into the core where it can undergo fusion. This lengthens the duration of the MS phase. The M dwarf stars will eventually end up as helium white dwarfs, supported by electron degeneracy pressure.

5.4.3 Evolution of Low Mass Stars

Here, we describe in some detail the evolution of a Sun-like star. Figure 5.4 illustrates the interior structure at some key phases of evolution, and Figure 5.5 illustrates the evolutionary track in the H-R diagram. Once the fusion (often called "burning") of hydrogen is over, the core starts contracting gravitationally. This contraction is accompanied by the initiation of hydrogen fusion in a surrounding shell that was previously unable to undergo fusion. The star becomes more luminous and the off-center nuclear burning also causes the outer layers to expand. The surface temperature drops during this initial post-main-sequence phase known as the **subgiant phase**. The outer layers are now cooler than before, and also become very opaque due to the high opacity of the H^- ion, a proton with *two* orbiting electrons that can form efficiently at the lower temperatures of an expanded envelope. In a hydrogen gas that contains free electrons that are generated from the easily ionized outer shells of metals, the reaction

$$\gamma + H^- \rightleftharpoons e^- + H \tag{5.32}$$

occurs in both directions. Photons with an energy as low as 0.75 eV are absorbed and emitted. Hence, a hydrogen gas can become both opaque and luminous. The enhanced opacity leads to a strong temperature gradient according to Equation (5.11), and the envelope becomes convective. The abundance of the H^- ion and its associated opacity peaks at a temperature of about 4000 K, since the metals become fully ionized at this temperature. At higher temperatures, no further electrons from metals are available and the environment also becomes too hot for the existence of a significant population of the loosely bound H^- ion. Conversely, at temperatures below 3000 K, the number of free electrons drops significantly and the abundance of H^- drops even more sharply. The gas is not very opaque at lower temperatures. In a remarkable result of stellar astronomy, Chushiro Hayashi (1920 - 2010) showed in the early 1960s (Hayashi, 1961, 1966) that stars with deep convective zones (e.g., PMS stars and giant branch stars) cannot get to the right (i.e., to lower surface temperature) of a nearly vertical line in the H-R diagram now known as the **Hayashi track**. The observed surface (photosphere) where the optical depth measured from outside is about unity, is always about 3000 K for these convective stars. Therefore, stellar evolution takes such stars along a path with nearly constant $T_e \approx 3000$ K. If the nuclear energy generation rate increases, a star will evolve vertically upward along the Hayashi track, increasing its luminosity but keeping T_e nearly constant; hence, the relation $L = 4\pi R^2 \sigma T_e^4$ means that the radius R increases dramatically. A star on this initial rise along the Hayashi track is on the **red giant branch**. A $1\,M_\odot$ star can expand to $R \approx 50\,R_\odot$ during this phase.

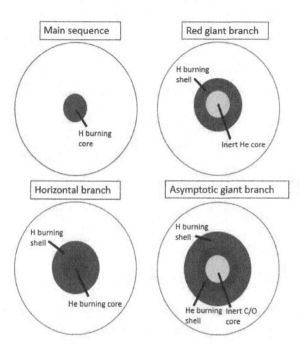

Figure 5.4 An illustration of the nuclear burning inside a Sun-like star at different evolutionary phases as labeled. The sizes of the stars and interior shells are not to scale. If the radius of a $1\,M_\odot$ star on the main sequence is $R_\odot$, the radius is $\approx 50\,R_\odot$ on the red giant branch, is $\approx 5\,R_\odot$ on the horizontal branch, and $\approx 300\,R_\odot$ on the asymptotic giant branch.

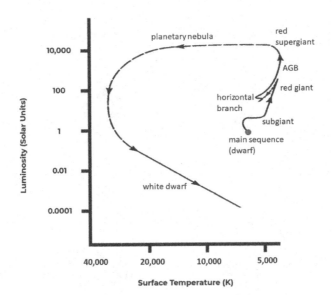

Figure 5.5 The evolutionary path of a 1 $M_\odot$ star. The dashed portion occurs very rapidly and is not well constrained by observations and theory. Adapted from calculations presented by Iben (1967).

The still contracting core eventually becomes hot enough to trigger helium burning. However, by this time, the electrons in the core have become degenerate. This means that the dependence of pressure on temperature is negligible. In such a case, nuclear reactions increase the temperature, thereby further increasing the reaction rate. There is no corresponding increase in pressure to create an envelope expansion. Therefore, helium burning occurs as a runaway process called the **helium flash**, a term coined by Leon Mestel (1927 - 2017). Eventually, the increasing temperature leads to the dominance of thermal pressure, and the core expands and a stable helium burning phase is reached, albeit at a lowered luminosity. This stable phase of core helium burning (into carbon and oxygen) and hydrogen shell burning is known as the **horizontal branch**. The branch is horizontal because stars of slightly different mass and composition and differing histories of mass loss (from the extended poorly bound envelope) during the red giant phase can settle to quite different locations along the branch. The star is now in a much more compact state than when it was in the red giant branch, and although there is a spread of possible values we can adopt a characteristic radius $\approx 5\,R_\odot$.

When the helium core burning (into carbon and oxygen) is completed, the core contracts again. Now, there is helium burning initiated in a shell just outside the core, and hydrogen burning occurs in a shell just outside of that. This double-shell burning **asymptotic giant branch** (AGB) is again characterized by envelope expansion, but of much greater extent than in the red giant branch. The envelope is again cool and convective, with significant opacity from H^-. The star once again follows a nearly vertical path upward in the H-R diagram but is characterized by much greater luminosity and larger radius than in the red giant branch. A star of $1\,M_\odot$ can reach luminosities $L \approx 10^3\,L_\odot$ and radius $R \approx 300\,R_\odot$. The Earth will in the future be engulfed by the Sun in its AGB phase, since the mean Earth–Sun distance 1 AU $\approx$ $200\,R_\odot$.

A star at its late AGB phase is a red **supergiant**, and during this phase, there is tremendous mass loss from the poorly bound outer envelope. This is aided by the sensitivity of helium shell burning to temperature, which can drive a series of episodes of expansion of contraction known as **thermal pulses**. Stars in this stage may exhibit significant changes in luminosity and are often categorized as **variable stars**. The mass that is lost by the star forms a surrounding shell, which if sufficiently illuminated from the stellar core, is known as a **planetary nebula**. This name carries the historical baggage of their early misidentification as planets due to their colorful gases that resemble the outer atmosphere of gas giant planets like Jupiter. Figure 5.6 shows a famous example of such an object, the Helix Nebula.

The hot core of the star that remains after the loss of the outer envelope is known as a **white dwarf**. The core is now dense enough that the effect of electron degeneracy pressure (discussed quantitatively in Chapter 7) dominates the thermal pressure of the non-degenerate ions. This degeneracy pressure is sufficient to prevent further contraction, and the temperature is not high enough for fusion of the carbon and oxygen nuclei. The surface energy loss by blackbody radiation continues, but this does not affect the dominant electron degeneracy pressure. Hence, the object slowly cools down while retaining its hydrostatic equilibrium for all time!

Figure 5.6 The Helix Nebula. A white dwarf at the center has earlier shed its outer layers, creating a glowing planetary nebula that is absorbing energy from the blackbody radiation of the white dwarf and re-emitting energy at characteristic wavelengths of Hα (red) and O III (blue-green). Credit: Adam Block, www.AdamBlockPhotos.com.

To summarize, we can describe the evolution of a low-mass ($\approx 1\,M_\odot$) star in terms of the following phases.

Zero-Age Main Sequence (ZAMS): In this phase of a star, hydrogen starts burning at the core.

Evolution on Main Sequence: Hydrogen at the core is fused to form helium, accounting for most of the star's nuclear burning lifetime.

Subgiant Phase: Hydrogen burning has ended, forming a helium core, whereupon helium burning begins in a surrounding shell and the envelope starts expanding while the core contracts.

Red Giant Branch: The envelope continues to expand during this phase. The luminosity increases and eventually the envelope becomes convective due to increasing opacity. The core is supported by electron degeneracy pressure.

Horizontal Branch: During this phase, helium burning is initiated in the core and starts with a flash due to degeneracy support, before the higher temperature makes the core nondegenerate and allows it to settle down to stable helium burning. Hydrogen in the shell continues to burn and the star settles into a stable analog of the main sequence.

Asymptotic Giant Branch: This is the stage at which all helium in the core is exhausted. Now, hydrogen and helium in surrounding shells start burning, in a process known as **double-shell burning**. This results in an expansion of the envelope and shrinking of the core.

Thermal Pulses (Variable Star): At this stage, the sensitivity of helium shell burning to temperature leads to cycles of expansion and contraction, so the star has a variable luminosity.

Planetary Nebula: The pulsations from the previous stage lead to a super-wind that ejects the stellar envelope. The hot core heats up and ionizes the ejected envelope.

White Dwarf: The core does not reach the ignition temperature of carbon; therefore, all thermonuclear reactions stop. The object remains in hydrostatic equilibrium due to electron degeneracy pressure and cools over time.

5.4.4 Evolution of High Mass Stars

The primary difference between the evolution of low and high mass stars is the core temperature. In stars with mass $\gtrsim 8M_\odot$, the temperatures are high enough to fuse the elements up to iron. Up to this element, nucleosynthesis produces more nuclear binding energy; therefore, there is an energy release. After hydrogen burning ceases, the subsequent phases occur quite rapidly. Each phase is shorter than the previous one, as the core contracts and the temperature rises, leading to each successive fuel burning at an expedited pace. For example, in a 25 $M_\odot$ star, the core carbon burning phase may last about 50 yr and the core neon and oxygen burning phases will each only be several months long. By the time that the core has fused elements up to iron, it is surrounded by an onion-like structure of shells that are burning different elements, as illustrated in Figure 5.7. The core and surrounding nuclear burning shells have a radius of only $\approx 10^{-2} R_\odot$ while the envelope has expanded to supergiant status with an outer radius $\approx 10^3 R_\odot$.

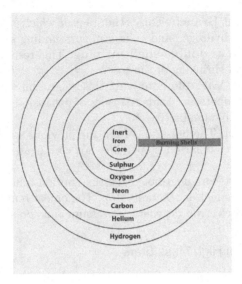

Figure 5.7 A schematic of the core of a very evolved high-mass star, just prior to a core collapse supernova. The innermost core is inert and composed of iron, with a series of outer shells that are burning different elements as labeled.

Since the outer layers do not have much time to respond to the rapid changes in the core, there is a steady drift to the right in the H-R diagram (Figure 5.8). Eventually, the core reaches a crisis point, the **iron catastrophe**. The star now has an iron core and exothermic nuclear reactions are no longer possible. Additional energy would be required to fuse iron into heavier elements. The conditions are ripe for a **core collapse supernova**.

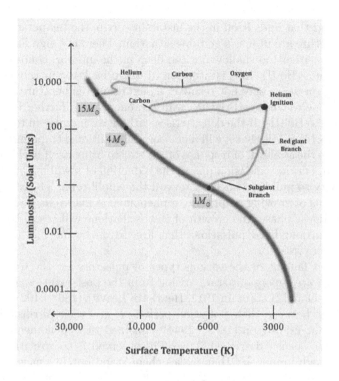

Figure 5.8 Partial evolutionary tracks for stars of two different masses. Points where the burning of various elements start are labeled.

5.4.5 Instability Strip

The instability strip has been mentioned earlier and shown in Figure 5.3. Stars that cross the instability strip during the course of their post-main-sequence evolution find themselves in a very special situation. In a typical star, the opacity κ decreases when the density increases. This can be understood from Equation (5.14), since compression will typically increase the temperature T as well as the density ρ, so the stronger dependence on inverse temperature means that κ will decrease. For a star in hydrostatic equilibrium, any fluctuation that drives up the density will result in a rebound due to the internal pressure, but this rebound will be damped by the effect of lowered opacity, allowing more photons to escape than usual, and reducing the restoring effect of radiation pressure. In the expansion phase, κ increases above the equilibrium value, thereby increasing the internal pressure resistance to gravity that is pulling the star back to the initial state. Hence, the return to equilibrium occurs more slowly than it would if κ had remain fixed. As a result, the usual response of a star to any hypothetical compression or expansion is a series of damped oscillations about the equilibrium state.

For a star that finds itself in the instability strip, the temperature and envelope structure are in just a suitable situation. There is a significant portion of the star, neither too shallow nor too deep in the interior, containing singly ionized helium (He II). In this situation, if we go back to our thought experiment, a compression may not result in a significant temperature increase, as the work done to compress the gas will feed energy into further ionization of the helium (to He III). If the density rises without an increase in temperature, Equation (5.14) shows that κ will increase. The additional trapping of photons will increase the strength of response of the star to bring itself back toward the equilibrium. On the expansion phase, the reduction of κ will increase the ability of gravity to pull the star back toward the equilibrium. The result is a set of oscillations of growing amplitude, a phenomenon known as **overstability**. In the nonlinear phase, the growth of the oscillations will cease, but the star can maintain long-lived pulsations that are driven by internal fluctuations, e.g., from convection.

The most famous of the various types of pulsators are the Cepheid variables, which are supergiant stars arising from the post-main-sequence evolution of type O and B stars. In 1912, Henrietta Leavitt (1868 - 1921) measured a correlation between their pulsation periods P and their intrinsic luminosities L. Arthur Eddington (1882 - 1944) explained this phenomenon in terms of the reverse opacity described above. The observed P-L correlation, coupled with their high luminosity that makes them stand out in images of distant galaxies, elevated Cepheids to being a preferred means of measuring the **distance scale** of the universe. A measurement of the apparent magnitude of a Cepheid coupled with the inferred absolute magnitude from the P-L relation yields a distance. Edwin Hubble (1889 - 1953) used the Cepheids to prove that there were galaxies outside of our own and to discover that distant galaxies are moving away from us: the **Hubble law**.

Figure 5.3 shows that even some white dwarfs can be found in the instability strip. Those white dwarfs have retained an outer layer of hydrogen and helium where the reverse opacity effect can occur.

5.5 SUMMARY

We conclude this chapter with some main points.

5.5.1 Stellar Structure

The theory is based on three basic principles.
Hydrostatic Equilibrium: A star is in mechanical equilibrium with the pressure at every level equal to the weight of a column of material per unit cross-sectional area on top.
Energy Transfer: Photons in the interior carry energy outward by random walking from regions of higher temperature to regions of lower temperature. If the luminosity required to be carried out is too large for this process, convection results.

Energy Generation: Energy release in the interior, through nuclear reactions, balances the outward release of energy through radiation or convection. If the nuclear source is inadequate, gravitational contraction must occur.

5.5.2 Stellar Evolution

Models of stellar evolution proceed through a solution of the fundamental equations of stellar structure, with small ongoing adjustments due to the energy generation. Here are some evolutionary highlights that you should remember.

1. The evolutionary history of a star depends on only two parameters: its mass M and its initial composition μ.

2. The pre-main-sequence phase is powered by gravitational contraction with no nuclear reactions.

3. $M \approx 0.075 M_\odot$ is the minimum mass for sustained nuclear reactions that allow an object to reach the main sequence.

4. A star is generally stable during nuclear burning. For example, an increase in ϵ leads to expansion that in turn lowers T_c and thereby decreases ϵ.

5. Very low mass stars ($M/M_\odot < 0.5$) have main-sequence lifetimes longer than the estimated age of the universe.

6. The Sun's main-sequence lifetime is about 10^{10} yr.

7. Very high mass stars have a main-sequence lifetime of only a few Myr.

8. A star gets hotter as it evolves, even though energy is continually radiated away. This is a gravitational effect.

9. The main-sequence phase is followed by shorter periods of burning successively heavier elements, proceeding as far as the star's pressure and temperature can allow. The star's outer envelope expands during this time.

10. Low mass stars eventually cease nuclear reactions and settle down to a degeneracy supported state, a white dwarf.

11. High mass stars can reach the iron catastrophe and end their lives with core collapse supernovae.

CHAPTER 6

Stellar and Planetary Systems

6.1 INTRODUCTION

About two-thirds of Sun-like stars are found in multiple systems. Most of these are binary systems, pairs of stars that orbit each other due to a mutual gravitational attraction. A small portion are in triple or quadruple systems. The probability of multiplicity is actually a strong function of stellar mass, with the lowest mass stars (the M stars) having a low probability ($\sim 20\%$) of being in multiple systems, and the highest mass stars (the O stars) having $> 70\%$ probability of being in a multiple system. Since Sun-like stars ($M \approx 0.7 - 1.3\, M_\odot$; spectral types F through mid-K) have about a 50% multiplicity fraction (see Figure 6.1), and most of the multiple systems are binaries, we can say that about two out of three Sun-like stars are in multiple systems. On the other hand, since at least 70% of all stars are M stars, we can also say that a majority of all stars are living alone, that is, without a stellar companion.

DOI: 10.1201/9781003215943-6

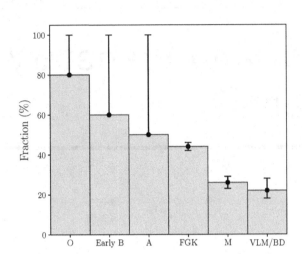

Figure 6.1 The multiplicity fraction of Pop I stars and brown dwarfs versus spectral type, measured by analysis of binary motions, based on data from Raghavan et al. (2010); Duchêne & Kraus (2013). Error bars cover the estimated range of uncertainty in each category. The categories of stellar type are O ($M \gtrsim 16\,M_{\odot}$), Early B ($M \approx 8 - 16\,M_{\odot}$), A ($M \approx 1.5 - 5\,M_{\odot}$), FGK ($M \approx 0.7 - 1.3\,M_{\odot}$), M ($M \approx 0.1 - 0.5\,M_{\odot}$), and Very Low Mass/Brown Dwarf, or VLM/BD ($M \lesssim 0.1\,M_{\odot}$).

Binary systems are crucial for our understanding of all stars and stellar evolution. This is because they enable a determination of the masses of stars in the first place. We measure masses here on Earth by measuring their acceleration due to a known force, or by the force they exert on other objects. This is hard to do for single stars in the depths of space, but binary pairs allow exactly this kind of measurement since we can measure their relative motion and combine with the basic Newtonian equations of motion and gravitation. Binary systems can also yield information about stellar radii, density, temperature, luminosity, and rotation rate.

The masses determined from binary system motions allow a compilation of the stellar mass–luminosity relationship for main sequence stars. It reveals a very steep dependence of luminosity on the stellar mass, as seen in Figure 6.2.

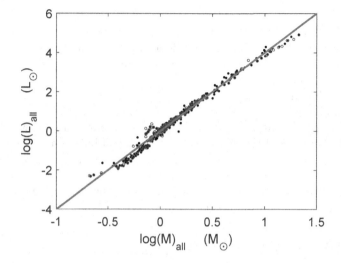

Figure 6.2 The luminosity L versus mass M of main sequence stars arising from analysis of binary systems. Data points are a compilation from several catalogs. The red solid line is a power law $L \propto M^4$. Credit: Wang & Zhong (2018), reproduced with permission from ESO.

It should be noted that $10^{-1} \lesssim M/M_\odot \lesssim 10^2$ but $10^{-4} \lesssim L/L_\odot \lesssim 10^6$! The range of luminosities is far greater then the range of masses. In this way, the highest mass stars with $L \approx 10^{5-6}\,L_\odot$ can have an outsized influence on their surroundings and even the Galaxy as a whole. The approximate relation

$$L \propto M^\alpha \qquad (6.1)$$

is usually fit in a piecewise form in the low mass, intermediate mass, and high mass regimes. Much of the data is available for intermediate mass Sun-like stars, and in this region, $\alpha = 4$ provides a very good fit. When fitting only the very low mass regime $M < 0.5\,M_\odot$, a value of about 2.5 can be adopted. When fitting a single power law for the entire mass range, an average value $\bar{\alpha} = 3.5$ is sometimes adopted.

It is worth pointing out that two-body systems are nature's most perfect and stable gravitationally bound configuration. Isaac Newton (1642 - 1727) found an analytic solution for the two-body problem, and the orbits are closed and can continue in a stable manner for perpetuity. This happy scenario is ruined by any additional bodies, or by tidal effects if the bodies have significant size in comparison to their separation. In multi-body systems, including triple systems or stars with surrounding planets, smaller mass bodies can eventually be ejected and the motions of the objects can eventually become chaotic. Triple-star systems will typically eject the lowest mass member or at least kick that object out to a very large orbit in relation to the orbit radius of the two primary objects. This is likely the explanation for the α Centauri system, the closest stellar system to Earth.

Binaries can be classified into the following main categories.

Visual binary: In this system, two stars are resolved as separate points that orbit one another.

Astrometric binary: In this system, only one star is seen, but it wobbles in the sky implying the presence of an unseen companion.

Spectroscopic binary: This is an unresolved binary pair for which we see a periodic variation of the Doppler shift of spectral lines that is called double lined if both spectra are observed or single lined if only one spectrum is observed.

Eclipsing binary: This is an unresolved binary that changes periodically in the total light emission due to the stars eclipsing each other.

6.2   ANALYSIS OF BINARIES

6.2.1   Visual and Astrometric Binaries

The Earth's atmosphere limits traditional ground-based optics to a resolution $\Delta\theta < 1''$, so visual and astrometric binaries are traditionally the ones with large separation, which implies a long period.

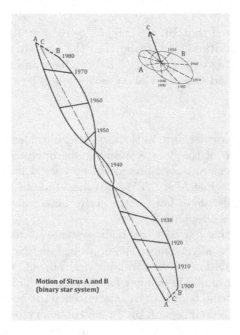

Figure 6.3   Observations of the motions of Sirius A and B, before and after the center of mass motion is subtracted out.

A famous example of an astrometric binary that is also a visual binary is the pair Sirius A and Sirius B; the latter is actually a white dwarf as discussed in Chapter 7. Figure 6.3 shows a time-sequence of observed positions of the two objects, both before and after the center of mass motion is subtracted out. Figure 6.4 shows a visual image of the binary pair. Sirius A (or Sirius) is the brightest star in the sky (aside from the Sun) and it was first established to be part of an astrometric binary in 1844 by Friedrich Bessel (1784 - 1846). He noticed the wobbling motion of Sirius A by examining many years of data of its proper motion, or motion in the plane of the sky relative to the background of stars. Sirius B was first detected optically in 1862 by Alvan Graham Clark (1832 - 1897).

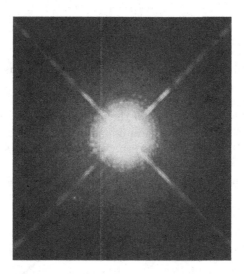

Figure 6.4 Image of Sirius A and Sirius B taken by the *Hubble Space Telescope*. The small dot to the lower left of the bright star Sirius A is Sirius B. Credit: NASA.

In order to analyze visual binaries, we observe their apparent (projected) elliptical orbit by first subtracting out the center of mass motion of the system. From the elliptical orbit, one can measure the angular size (α) of the elliptical semimajor axis a of the relative orbit between the two stars, so $\alpha = a/d$ where d is the distance to the binary. Note that a is also the average separation between the stars. We can also measure the period P of the orbit. We can use astronomers units to say that

$$a(\text{AU}) = \alpha('') \, d(\text{pc}), \tag{6.2}$$

where a is measured in AU, α in arcseconds ($''$), and d in pc. This relation employs the definition of 1 pc, i.e., that it is the distance at which a size of 1 AU has an angular extent of $1''$.

Once we know the period we can also employ Kepler's third law:

$$P^2 = \frac{4\pi^2 a^3}{G(m_1 + m_2)} \Rightarrow m_1 + m_2 = \frac{4\pi^2}{G} \frac{a^3}{P^2}. \tag{6.3}$$

Here, m_1 and m_2 are the masses of the two stars. We also note that the elliptical semimajor axis $a = a_1 + a_2$, where a_1 and a_2 are the semimajor axes of the individual elliptical paths traversed by each star with respect to the center of mass. The center of mass moves in a straight path with respect to the background stars, so its motion and positions are easily determined. Since this proper motion reveals a_1 and a_2, we use the definition of center of mass,

$$\frac{m_1}{m_2} = \frac{a_2}{a_1}, \tag{6.4}$$

to obtain a second equation. Together, equations (6.3) and (6.4) can be solved to yield individual values for m_1 and m_2. Figure 6.5 illustrates the case of circular orbits.

For completeness, we note that the values of a, a_1, and a_2 can only be determined accurately by correcting for the inclination of the orbit plane relative to the plane of sky. These planes do not generally coincide, but the relative angle i (see Figure 6.6) can be discerned by careful analysis of the displacement of the primary star from the apparent focus of the projected elliptical orbit. There is a measurable displacement when $i \neq 0°$. Details are beyond the scope of the presentation here.

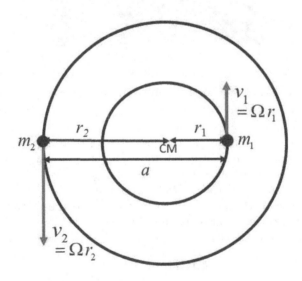

Figure 6.5 An illustration of two planets in a circular orbit.

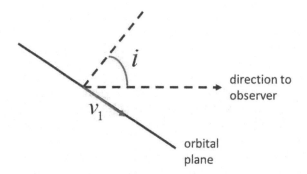

Figure 6.6 An illustration of the inclination angle i in the analysis of binary orbits. If a star has a velocity v_1 in the orbit plane, then an observer viewing from a line of sight characterized by the angle i will measure a line-of-sight velocity $K_1 = v_1 \cos(90° - i) = v_1 \sin i$.

6.2.2 Spectroscopic Binaries

These are binary systems that cannot be resolved as individual objects by a telescope, but the spectra reveal periodic velocity motions along the line of sight. The spectrum from the secondary star may be too faint to be detected, in which case the spectroscopic binary is known as **single lined**. If spectra from both stars are detected, they undergo inverse Doppler shifts, so that one set of lines is redshifted when another is blueshifted. These cases are known as **double lined**.

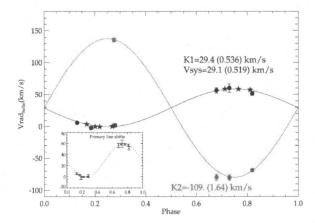

Figure 6.7 The variation of velocity inferred from the Doppler shift of spectral lines in a double-lined spectroscopic binary. The orbits are circular and the black and red solid lines are the best-fitting sinusoidal functions. The binary pair have a systematic motion of $v_{\text{sys}} = 29.1$ km s^{-1} relative to Earth. Once this is subtracted out, the line of sight motions of the two stars have maximum amplitude K_1 and K_2. Uncertainties in values are given in parentheses. Credit: Nefs et al. (2013), reproduced with permission from OUP.

There is a direct relation between the Doppler shift of spectral lines and radial (line-of-sight) velocity. For a shift of wavelength $\Delta\lambda$, the corresponding line of sight velocity is $v \sin i = c\Delta\lambda/\lambda$, where v is the magnitude of orbit velocity and i is the inclination of the orbit plane relative to the plane of the sky. For the two stars in a double-lined system, we measure both $K_1 = v_1 \sin i$ and $K_2 = v_2 \sin i$ (see for example Figure 6.7). If the system is single lined, we can only detect the spectral lines from one of the two stars. Here, we analyze these binaries by applying the laws of orbital motion, using the measured K_1, K_2, and the period P. We consider the following three cases.

6.2.2.1 Double Lined, $i = 90°$, circular orbit

This is an illustrative case, keeping in mind that eclipsing binaries always have i close to $90°$; otherwise, they would not be eclipsing. Furthermore, they are in close orbits, and therefore, the tidal effects are very effective in circularizing the orbits. For $i = 90°$ (edge on) and a circular orbit, we observe a sinusoidal velocity curve (non-circular orbits would yield non-sinusoidal oscillations in velocity). Therefore we observe v_1, v_2, and P as follows:

$$P = \frac{2\pi r_1}{v_1} = \frac{2\pi r_2}{v_2},$$

(6.5)

where r_1 and r_2 are the radii of the circular orbits of each object, measured relative to the position of the center of mass of the system. The separation between the stars is

$$a = r_1 + r_2 = \frac{P}{2\pi}(v_1 + v_2).$$

(6.6)

Furthermore,

$$\frac{m_2}{m_1} = \frac{r_1}{r_2} = \frac{v_1}{v_2}.$$

(6.7)

We can combine the above two equations with Kepler's third law to get

$$m_1 + m_2 = \frac{4\pi^2}{G}\frac{a^3}{P^2} = \frac{P}{2\pi G}(v_1 + v_2)^3.$$

(6.8)

Equations (6.7) and (6.8) can be combined to solve for m_1 and m_2.

6.2.2.2 Double Lined, arbitrary i

For a double-lined binary with arbitrary i, we now observe P, $K_1 = v_1 \sin i$ and $K_2 = v_2 \sin i$. Then

$$\frac{m_2}{m_1} = \frac{r_1}{r_2} = \frac{v_1}{v_2} = \frac{K_1}{K_2}.$$

(6.9)

Note that $\sin i$ cancels out. Furthermore,

$$P = \frac{2\pi r_1}{v_1} = \frac{2\pi r_1 \sin i}{K_1} = \frac{2\pi r_2 \sin i}{K_2},$$

(6.10)

which implies that

$$r_1 = \frac{K_1 P}{2\pi \sin i}, r_2 = \frac{K_2 P}{2\pi \sin i}.$$

(6.11)

Since

$$a = r_1 + r_2 = \frac{P}{2\pi \sin i}(K_1 + K_2),$$

(6.12)

we find that

$$m_1 + m_2 = \frac{4\pi^2}{G}\frac{a^3}{P^2} = \frac{P}{2\pi G}\frac{(K_1 + K_2)^3}{\sin^3 i}.$$

(6.13)

Once we have measured K_1 and K_2, we can combine equations (6.9) and (6.13) to solve for $m_1 \sin^3 i$ and $m_2 \sin^3 i$. Note that equation (6.12) also allows a determination of $a \sin i$ but not a itself.

6.2.2.3 Single Lined, arbitrary i

For a single-lined binary with arbitrary i, we only observe P and $K_1 = v_1 \sin i$. Looking again at the case of a circular orbit, we have

$$r_1 = \frac{v_1 P}{2\pi} = \frac{K_1 P}{2\pi \sin i}.$$

(6.14)

Then using the definition of center of mass, we find

$$a = r_1 + r_2 = r_1(1 + m_1/m_2) = \frac{K_1 P}{2\pi \sin i}(1 + m_1/m_2). \qquad (6.15)$$

Using Kepler's third law we find

$$m_1 + m_2 = \frac{4\pi^2}{G}\frac{a^3}{P^2} = \frac{K_1^3 P}{2\pi G \sin^3 i}(1 + m_1/m_2)^3. \qquad (6.16)$$

This leads to

$$m_2^3 \sin^3 i = \frac{K_1^3 P}{2\pi G}(m_1 + m_2)^2. \qquad (6.17)$$

If $m_1 \gg m_2$, then

$$m_2 \sin i = \frac{K_1 P^{1/3}}{(2\pi G)^{1/3}} m_1^{2/3}. \qquad (6.18)$$

If we know the mass m_1 of the primary star, then this yields a lower limit to m_2, the mass of the unseen companion. This is because $m_2 \geq m_2 \sin i$. This method has proven very fruitful to find exoplanets.

6.2.3 Eclipsing Binaries

These are binaries that have a relatively close orbit and the inclination angle i is close enough to 90° that we see repeating eclipses, where one star passes in front of another. Since the stars generally have two different temperatures, the brightness decrease is greater during a primary eclipse, when the star with greater temperature T_{hot} passes behind the star with lesser temperature T_{cool}. Figure 6.8 illustrates the sequence of eclipses.

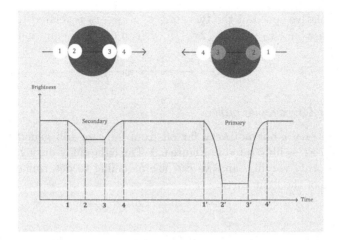

Figure 6.8 An illustration of the observed brightness during eclipses for an eclipsing binary whose orbit plane is perpendicular to the plane of the sky, or $i = 90°$. During the secondary eclipse, the smaller star, which is assumed to be the hotter of the two, passes in front of the larger one. During the primary eclipse, which occurs one half orbit period later, the smaller star passes behind the larger one. For each eclipse, the four numbered contact times characterize distinct events.

6.2.3.1 $i = 90°$, circular orbit

Here, we demonstrate the analysis when $i = 90°$ and the stars are in a circular orbit. Let R_s and R_l be the radii of the small and large stars, respectively. Figure 6.8 illustrates key moments during the eclipses. The primary eclipse occurs when the smaller star (assumed to be the hotter of the two) passes behind the larger star, and the secondary eclipse occurs when the larger (and cooler) star passes behind the smaller star. The labeled times (primed in some cases) represent the following events: t_1 is the time of first contact, when the eclipse begins; t_2 is the second contact, when the brightness minimum is reached; t_3 is the third contact, when the smaller star begins to leave the disk of the larger star; t_4 is the fourth contact, when the eclipse ends. At each of these four points, one limb of a star is tangent to one limb of the other star.

Considering the secondary eclipse for example, we can write the following, employing v as the relative speed of the two objects in a circular orbit:

$$2R_s = v(t_2 - t_1) = v(t_4 - t_3), \tag{6.19}$$

$$2(R_s + R_l) = v(t_4 - t_1). \tag{6.20}$$

Since the relative speed of the two stars $v = v_1 + v_2$ is related to their total separation $a = r_1 + r_2$ by $P = (2\pi a)/v$, we can combine equations to obtain

$$\frac{R_s}{a} = \pi \frac{t_2 - t_1}{P}, \quad \frac{R_l}{a} = \pi \frac{t_4 - t_2}{P}. \tag{6.21}$$

6.2.3.2 $i \neq 90°$, circular orbit

For an arbitrary i but not very far off from $90°$, we will generally observe partial eclipses, as illustrated in Figure 6.9. The light curve during the eclipses is now smoothly varying, and we can use modeling to determine i from the exact shape of the light curve.

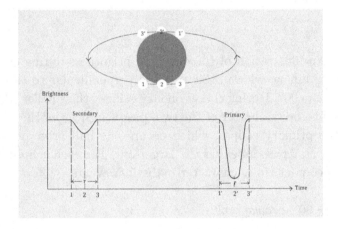

Figure 6.9   An illustration of an eclipsing binary when the orbit is slightly tilted, resulting in partial eclipses.

Figure 6.9 illustrates key moments during the partial eclipses. The primary eclipse occurs when the smaller star (assumed to be the hotter of the two) passes behind the larger star, and the secondary eclipse occurs when the larger (and cooler) star passes behind the smaller star. The labeled times (primed in some cases) represent the following events: t_1 is the time of first contact, when the eclipse begins; t_2 is the second contact, when the brightness minimum is reached; and t_3 is the third contact, when the eclipse ends.

The partial eclipse binaries also provide a means to measure the relative temperature of the two stars. Since the flux F from each star is proportional to T^4, and the area eclipsed is the same in the primary and secondary eclipses (see Figure 6.9), we see that

$$\frac{\text{depth of primary}}{\text{depth of secondary}} = \frac{F_{\text{Primary}} \times \text{area eclipsed}}{F_{\text{Secondary}} \times \text{area eclipsed}} = \left(\frac{T_{\text{hot}}}{T_{\text{cool}}}\right)^4. \tag{6.22}$$

6.2.3.3 Final Thoughts

A fortunate thing about eclipsing binaries is that they are generally also spectroscopic binaries. This means that they complete the final piece in our understanding of stars by allowing us to combine data sets and reliably measure stellar masses and radii, among other properties.

Recall that the analysis of spectroscopic binaries yields values for

$$a \sin i, \; m_1 \sin^3 i, \; m_2 \sin^3 i, \qquad (6.23)$$

while eclipsing binary data for the same system yields

$$\frac{R_s}{a}, \; \frac{R_l}{a}, \; i. \qquad (6.24)$$

Combining the information, we extract a treasury of data, yielding values for each of

$$m_1, \; m_2, \; a, \; R_s, \; R_l \; ! \qquad (6.25)$$

The above method leads to the stellar mass–luminosity relation (Figure 6.2) as well as the stellar mass-radius relation, two of the great achievements of observational stellar astronomy. These results help to develop, constrain, and verify theories of stellar structure and evolution.

6.2.4 Close Binaries

Let us consider a binary system with short separation a and period P and transform to a frame where the stars are stationary, i.e., a frame rotating at a rate $\Omega = 2\pi/P$. In this frame,

$$g_{\text{eff}} = g_1 + g_2 - \Omega^2 r, \qquad (6.26)$$

where r is the distance to the center of mass, and g_1 and g_2 are the gravitational fields of the two stars. A sideways figure-eight defines the two regions where g_1 and g_2 dominate, respectively. These regions are called the Roche lobes.

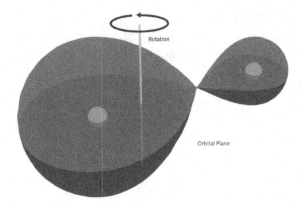

Figure 6.10 An illustration of the Roche lobes of two stars in a binary orbit. The star on the left has three times the mass of the star on the right. The system rotates about the center of mass, situated at the base of the rotation axis vector.

Close binaries are classified into three broad categories that depend on the relation of each star with its Roche lobe. These are as following.

Detached system: Both stars are smaller than their respective Roche lobes. This is illustrated in Figure 6.10.

Semi-detached system: One star fills its Roche lobe as a result of post-main-sequence evolution and mass flows to the companion. The angular momentum of the incoming material will lead to the formation of an accretion disk around the companion. This is illustrated in Figure 6.11

Contact system: Both stars fill their Roche lobes. The system is shrouded by a common envelope of material.

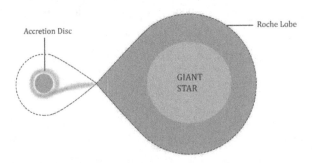

Figure 6.11 Illustration of a semi-detached binary system.

6.3 OBSERVING THE EXOPLANETS

All of the classically developed techniques that were developed to detect stellar companions can now be used to detect the much fainter planetary companions. What is needed is greater spectral resolution or more sensitive photometry (measurement of light). Since 1995, these techniques have been used to observe an ever increasing number of exoplanets, i.e., planets that orbit stars other than the Sun. Many brown dwarfs have also been detected through these techniques.

In exoplanetary science, the term **radial velocity technique** is used to describe the methods used for single lined spectroscopic binaries, while the term **transit method** is used to describe the methods used for eclipsing binaries. The term **direct imaging** is equated to the concept of visual binaries. Finally, the technique of **microlensing**, the bending of light from a background source by the gravitational field of an unseen planet, has also been used to detect some exoplanets.

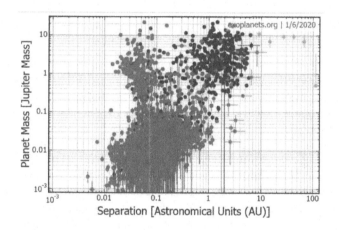

Figure 6.12 The estimated masses m (or $m \sin i$) and orbit radii of confirmed exoplanets, as of early 2020. Each dot represents an individual exoplanet detection, and the color denotes the discovery technique: radial velocity (blue), transit (red), microlensing (green), and direct imaging (yellow). The figure was generated at http://exoplanets.org.

The data in Figure 6.12 should not be considered to be a complete sample, i.e., a representative sample of the full exoplanet population. This is because all the detection methods have a **selection bias**. This means that each technique is better able to detect a certain subclass of exoplanets, dependent on their mass and position relative to their host star, or on the properties of the host star itself. The radial velocity technique is biased toward detecting stellar wobbles of $\sim 1 \ \text{m s}^{-1}$ or greater, given current technology.

To detect an Earth-like planet around a Sun-like star at a distance $r_\oplus = 1$ AU, requires greater precision. To see this, we make a quick estimate of the Sun's wobble that is induced by Earth. First, we can estimate the Earth's orbit velocity $v_\oplus$ assuming a circular orbit and that the Sun is close to being in a fixed position near the center of mass of the system. Then

$$\frac{v_\oplus^2}{r_\oplus} = \frac{GM_\odot}{r_\oplus^2} \Rightarrow v = \sqrt{\frac{GM_\odot}{r_\oplus}}. \tag{6.27}$$

Now, we account for the velocity of the Sun's motion $v_\odot$ by using the principle of conservation of momentum relative to the center of mass:

$$M_\odot\, v_\odot + M_\oplus\, v_\oplus = 0. \tag{6.28}$$

Therefore, the magnitude of the Sun's motion relative the center of mass of the Earth–Sun system (note this is a hypothetical value we are calculating assuming that the Earth is the Sun's only companion) would be

$$|v_\odot| = \frac{M_\oplus}{M_\odot} \sqrt{\frac{GM_\odot}{r_\oplus}} \approx 0.1 \text{ m s}^{-1}. \tag{6.29}$$

A similar calculation using Jupiter's mass and orbit radius shows that it induces a solar motion of ≈ 12 m s^{-1}. A perceptive alien that measures our Sun's wobble could however separate out (after long-time monitoring) the different components of the Sun's composite motion by harmonic analysis, given the different periods of the motions induced by different orbiting planets.

The small motions induced by Earth-like (or **terrestrial**) planets at ~ 1 AU separation from Sun-like stars means that they remain generally out of reach of detection by the radial velocity technique. This bias can be seen in Figure 6.12 by the distribution of blue dots, which is dominated by high-mass (or **gas giant**) planets, some of which are at extremely close separation ~ 0.05 AU. This class of close-orbiting giant planets have been given the name **hot Jupiters**, since they have Jupiter-class masses and face significant heating from their nearby host star. They were the first type of extrasolar planet to be detected, by Mayor & Queloz (1995), since they introduce a relatively large and more easily detectable wobble in their host star. Their locations remain mysterious, since all planet formation scenarios require the accumulation of such large masses to occur at large radii, from several AU to tens of AU. The close radii of hot Jupiters are usually attributed to an inward migration process due to interactions of the planet with their host disks during the early evolutionary phase. But then why would the migration stop just before the planet plunges entirely into the host star? One explanation is that ≈ 0.05 pc is the typical distance at which the pressure of the stellar magnetic field will truncate the disk, leaving effectively an inner gap in the disk. The inward migration, if driven by interaction with disk gas, would then stop at this radius (Lin et al., 1996).

The transit method was revolutionized for exoplanets by the *Kepler Space Telescope*, which during its main mission from 2009 to 2013 detected over 4000 candidate exoplanets, with over 2000 being confirmed. The confirmation process involves observing at least one full cycle of the orbit, including at least two full transits, and/or some other independent method of verification. The transit method, like the radial velocity method, is biased toward short period binaries that have an orbit plane that is close to edge-on from our line-of-sight. Kepler was able to measure light curves for over 500,000 stars in the region of the Cygnus constellation. The Kepler data haul revealed a wide diversity of planet types. Besides finding gas giant and terrestrial planets, it discovered a whole new class of planets that we now call **super-Earths**. These are planets that have masses in between that of Earth and Neptune. Some of these are also in the **habitable zones**, meaning the regions where the host star radiation is sufficient (extreme atmospheric conditions aside) to allow the existence of liquid water. The Kepler telescope's observations showed that super-Earths are abundant in our Galaxy. It is an open question whether life can exist on super-Earths if they are in the habitable zone. Another new class of planets are the so-called **super-puffs**. These are low-mass gaseous planets with radii similar to that of Jupiter but considerably sub-Jupiter masses, $< 10\,M_{\oplus}$. First detected in the system known as Kepler-51a (Masuda, 2014), there are up to 20 candidate super-puffs as of 2020, and these objects pose a challenge to standard theories of planet formation. The core accretion hypothesis generally requires the formation of a rocky core of at least $10\,M_{\oplus}$ before a gaseous envelope of hydrogen and helium can be captured from the surrounding nebula.

Extrapolation of Kepler data implies that there are approximately 10 billion terrestrial planets in our Galaxy that are in their host star's habitable zones, meaning that the fraction of stars with habitable Earth-like planets is at least ~ 0.1. When considering only Sun-like stars as hosts, the fraction with rocky planets that can sustain liquid water is estimated to be even higher, ~ 0.5 (Bryson et al., 2021). There is still much more to learn when it comes to exoplanets, and the pace of discovery is increasing rapidly, with many new missions underway or in planning. NASA's *Transiting Exoplanet Survey Satellite* (TESS, launched in 2018) is using the transit method to look for planets around nearby bright stars with enough precision to enable an analysis of the composition of their atmospheres by identifying molecular bands. The *James Webb Space Telescope* (JWST, set to launch in 2021) is expected to discover many faint exoplanets and separate out the spectral features of the planetary atmospheres. In doing so, it will search for molecules like oxygen and methane that are sustained in the Earth's atmosphere by biological processes. The hunt for such **biosignatures** as evidence for extraterrestrial life is a key goal of exoplanet studies.

A more complete sample of exoplanet types can come from long-time monitoring, which is needed for the radial velocity or transit method to identify wide-orbit exoplanets with their long periods. Furthermore, the increasing power of other techniques like direct imaging and microlensing can add to the exoplanetary census.

CHAPTER 7

Stellar Remnants

7.1 INTRODUCTION

During the main sequence phase, stars evolve due to a continual competition between the constraints of thermal equilibrium and mechanical equilibrium. As the astrophysicist Frank Shu has pointed out (Shu, 1982), there is a fundamental tension between the laws of thermodynamics and Newton's laws of gravitation and motion. The loss of energy by a star through radiation to its much cooler surroundings leads to gravitational contraction and raises the temperature of the core. A star can thus *increase* its internal temperature as a consequence of energy loss to its surrounding, earning itself a negative specific heat. Eventually, nuclear reactions cease and there is not enough internal thermal pressure to support hydrostatic equilibrium. The final outcome of such a predicament is one of the following.

White Dwarf: Electron degeneracy pressure (a quantum mechanical effect) supports the star.

Neutron Star: Neutron degeneracy pressure (also a quantum mechanical effect) supports the star.

Black Hole: Gravity wins out!

7.2 ORIGIN OF DEGENERACY PRESSURE

Can one observe quantum effects on a macroscopic scale? A white dwarf is exactly such an example. The effect originates from the **exclusion principle** introduced by Wolfgang Pauli (1900 - 1958), which states that no two fermions (particles with spin quantum numbers that have half-integer values; examples are electrons, protons, and neutrons) can occupy the same quantum state, and the **uncertainty principle**

$$\Delta x \, \Delta p_x \gtrsim h \qquad (7.1)$$

developed by Werner Heisenberg (1901 - 1976). Here, Δx is the positional uncertainty, or spread in values of the particle position x, and Δp_x is the

DOI: 10.1201/9781003215943-7

spread in values of momentum in the x-direction. The above equation implies that when quantum motions dominate,

$$p_x \sim \Delta p_x \gtrsim \frac{h}{\Delta x}. \tag{7.2}$$

We can infer from the above equations that a high density will cause large random motions, which in turn would increase the pressure of the system. Degeneracy pressure can exceed kinetic pressure ($P = nkT$) in very dense objects. One of the most interesting facets of degeneracy pressure is that it does not depend on temperature. Another way of thinking about degeneracy is that it sets in when the particle spacing Δx becomes small enough to be comparable to the **de Broglie wavelength** $\lambda_D \equiv h/p_x$ of the particles. Since electrons have much less mass than the nuclei in an ionized plasma, they have a much longer de Broglie wavelength, and become degenerate at a much lower density than the nuclei or nucleons.

7.2.1 The Degeneracy Parameter

In the classical limit, particle motions are described by the Maxwell–Boltzmann distribution, and particles of mass m have a peak momentum value $p_{\max} = (2mkT)^{1/2}$, corresponding to an energy $\epsilon = p_{\max}^2/(2m) = kT$. However, if the density n is sufficiently high, and/or the temperature T sufficiently low, momentum states get filled in accordance with the constraints of the uncertainty and exclusion principles. In the fully degenerate limit, fermions cannot coexist in the same momentum states (actually quantum mechanics allows two fermions of opposite spin within each momentum state), so degenerate particles fill up successively larger momentum states up to some maximum value p_F, the Fermi momentum, which is a function of the mean particle spacing Δx, and therefore the number density $n = (\Delta x)^{-3}$. From equation (7.2) we can infer that $p_F \propto h n^{1/3}$. An exact calculation yields

$$p_F = \hbar \left(3\pi^2 n \right)^{1/3}. \tag{7.3}$$

From the above relation, we may also define the Fermi energy, in the nonrelativistic limit ($p_F \ll mc$) as

$$\epsilon_F = \frac{p_F^2}{2m} = \frac{\hbar^2}{2m} \left(3\pi^2 n \right)^{2/3}, \tag{7.4}$$

and in the relativistic limit ($p_F \sim c$) as

$$\epsilon_F = p_F c = \hbar c \left(3\pi^2 n \right)^{1/3}. \tag{7.5}$$

An important consideration of the degree of degeneracy in any system is to calculate the dimensionless ratio $\mathcal{D} \equiv \epsilon_F/(kT)$. If $\mathcal{D} \ll 1$, the system is classical, whereas when $\mathcal{D} \gtrsim 1$, the system exhibits significant degeneracy pressure.

The (partial or fully) degenerate gas momentum distribution is described by the **Fermi-Dirac** statistics. For the remainder of this chapter, we will work in the fully degenerate limit. One can verify that $\mathcal{D} \gg 1$ for electrons in the core of a white dwarf and for neutrons in the interior of a neutron star.

7.3   WHITE DWARFS

White dwarfs are stellar remnants consisting of an exposed stellar core composed of carbon and oxygen nuclei and their associated electrons. We know from Chapter 4 that the internal pressure of a self-gravitating object of mass M and radius R is

$$P \sim \frac{GM^2}{R^4}. \tag{7.6}$$

In a situation of equilibrium, this must be balanced by the internal electron degeneracy pressure. The pressure due to microscopic particles can be understood through an argument of momentum transfer. For a surface oriented perpendicular to the x-direction, the pressure is the x-momentum carried across the surface per unit time per unit area:

$$P = n v_x p_x , \tag{7.7}$$

where n is the number density, and v_x and p_x are the velocity and momentum of particles in the x-direction, respectively. In practice, we take an ensemble average of the product $v_x p_x$ over the distribution function. In the nonrelativistic ($v \ll c$) case, $v_x = p_x/m$ for particles of mass m. Next, using equation (7.2) and mass density $\rho = mn \propto \Delta x^{-3}$, we obtain the nonrelativistic degenerate electron pressure equation

$$P = n v_x p_x \propto \rho^{5/3} \propto \left(\frac{M}{R^3} \right)^{5/3}. \tag{7.8}$$

Combining equations (7.6) and (7.8) we obtain

$$R \propto M^{-1/3} . \tag{7.9}$$

A more exact calculation yields the radius-mass relation

$$R = 0.114 \frac{h}{G m_e m_p^{5/3}} \left(\frac{Z}{A} \right)^{5/3} M^{-1/3} , \tag{7.10}$$

where m_e is the electron mass, m_p is the proton mass, Z is the atomic number of nuclei in the white dwarf, and A is the atomic mass of the nuclei. To derive this relation, one has to utilize the concept of overall charge neutrality, so that $n_e = Z n_i$, in which n_e and n_i are the number density of electrons and ions (atomic nuclei), respectively, and a mass density $\rho \simeq A m_p$ that is contributed by the ions. Here, we have also used the fact that the neutron mass m_n and proton mass m_p are nearly equal.

The white dwarf radius-mass relation means that the more massive a white dwarf, the *smaller* its radius. The inverse power relationship between radius and mass is quite unlike any object in our everyday experience, reflecting the extreme strength of gravity and the need for increasing density to achieve a balance with gravity as the total mass increases. This radius-mass relation is also markedly different from that of planets and main-sequence stars. The inverse relationship even hints that there may be a maximum possible mass of a white dwarf.

7.3.1   Limiting Mass of a White Dwarf

Subrahmanyan Chandrasekhar (1910–1995) realized that there is a loss of equilibrium at extremely high densities, when electrons become relativistic ($v \sim c$). The relativistic electron degeneracy pressure is

$$P = nv_x p_x \sim ncp_x \propto \rho^{4/3} \, . \tag{7.11}$$

Equating internal pressure with degeneracy pressure, as done earlier in the nonrelativistic case, now yields a maximum possible mass

$$M \propto \left(\frac{hc}{G} \right)^{3/2} \left(\frac{Z}{Am_p} \right)^2 \, . \tag{7.12}$$

A more exact calculation yields the Chandrasekhar limiting mass

$$M_{\mathrm{Ch}} = 0.20 \left(\frac{hc}{G} \right)^{3/2} \left(\frac{Z}{Am_p} \right)^2 \, . \tag{7.13}$$

For $Z/A = 0.5$, as for carbon and oxygen atoms that comprise white dwarfs, we get $M_{\mathrm{Ch}} = 1.40 \, M_\odot$.

7.3.2   Observations of White Dwarfs

A white dwarf of mass $M \approx 1 M_\odot$ has a radius $R \approx 10^{-2} R_\odot$, which implies an average density $\rho = 10^6 \, \mathrm{g\,cm^{-3}}$. How can we observe such an object? A famous example is Sirius B, the binary companion to the bright star Sirius. The two objects are in a binary orbit and their motion in the sky can be interpreted using Kepler's third law to yield a mass for Sirius B of $1.0 \, M_\odot$. The observed luminosity of Sirius B is $L = 3 \times 10^{-2} L_\odot$, and its temperature is $T_e = 25,000$ K. These yield a very small radius $9 \times 10^{-3} R_\odot$. Using the above values, we infer that the average density of the object is $2 \times 10^6 \, \mathrm{g\,cm^{-3}}$. Evidently, this object is a white dwarf.

7.3.3   Luminosity of a White Dwarf

Visible white dwarfs have a residual thermal energy that is slowly radiated away through blackbody radiation from the surface. The thermal energy is contained mainly in the ions (the atomic nuclei), rather than electrons, due to their greater mass. However, the electron degeneracy contains much more energy and provides the support against gravity. So, a white dwarf can be cooling but not contracting. Given enough time, it comes into thermal equilibrium with its surroundings and becomes a **black dwarf**.

Could a large population of faint white dwarfs or black dwarfs explain the hypothesized dark matter that is invoked to explain observations of galaxies and the large scale structure of the universe? Currently the answer is *no*, as various measurements have failed to find evidence for a widespread population of faint white dwarfs or black dwarfs.

7.4 SUPERNOVAE

Supernovae (SNe) have been discussed in previous Chapters on the Interstellar Medium and Stellar Evolution, but we revisit them here in the context of star death for two reasons. First, the core-collapse SNe that occur at the end of a massive star's life are thought to lead to the formation of neutron stars, the subject of the next section. Also, other SNe are thought to form when white dwarfs have accreted additional material from a binary companion and exceeded their Chandrasekhar limit. We review the different types of SNe here. First, the observational classification.

Type II: Observe strong hydrogen emission lines in the ejected material.

Type I: No hydrogen emission lines in spectrum, with the following subcategories.

    Type Ia: Remarkably uniform spectra and light curves.

    Type Ib: Less luminous than Type Ia and no silicon absorption line.

    Type Ic: Like Type Ib but no helium lines either.

The theoretical interpretations of these categories are as follows.

Type II: Core-collapse supernova from the end state of a massive star.

Type Ia: Explosion of a white dwarf that has accreted mass from a binary companion.

Type Ib: Core-collapse supernova from a massive star that has lost its hydrogen envelope.

Type Ic: Core-collapse supernova from a massive star that lost both hydrogen and helium envelopes.

One of the least understood features of a core-collapse supernova is exactly how a catastrophic collapse is converted into an explosive motion of the envelope. This is a frontier area of astrophysical research and the explanations involve some or all of the following: 1. a rebound effect when the infalling envelope encounters the very stiff neutron star that is forming at the center; 2. radiation pressure on the envelope; 3. neutrino absorption and scattering in the envelope; and 4. explosive nuclear reactions in the envelope.

The progenitors of the type Ib/c SNe may be the **Wolf–Rayet stars**, which are O-type stars at a very late stage of evolution that exhibit broad emission lines of ionized helium and/or other heavier ionized species. These stars have lost their hydrogen envelope and typically have an exposed helium core. They drive powerful winds at speeds of up to 3000 km s^{-1} and mass loss rates up to $10\,M_\odot$ Myr^{-1}. Given the mass loss rate, this is clearly a very short-lived phase, and the core is very close to the conditions for a core-collapse supernova.

The type Ia SNe occupy a very special place in astronomy. They show a remarkable uniformity in their spectra and their luminosity evolution (called **light curves** by astronomers). The working model for their origin is a white dwarf that is in a binary system and collecting matter from a companion that has exceeded the size of its Roche lobe. When the mass just exceeds M_{Ch}, an explosive fusion of carbon in the degenerate white dwarf can take place. Since the initial conditions for the SNe Ia are almost identical in each case, the result is a very small dispersion of values of the peak luminosity and decay time in the light curves. Even the observed dispersion can be accounted for as there is a known correlation between the peak luminosity and the decay time. Hence, the peak brightness of a type Ia supernova can be estimated even if the first measurement is made after the time of peak luminosity.

The SNe of type Ia are a major gift to astronomers, representing **standard candles** of known intrinsic brightness. Although only about two type Ia SNe are expected to take place in a typical galaxy within a millennium, the vastness of the observable universe means that one of these explosions occur somewhere in the sky every few seconds! If we measure the flux f at Earth and know the intrinsic luminosity L, we can estimate the distance to the supernova using a more sophisticated version of their relationship than we used in the context of parallax and the magnitude system; one that takes into account the stretching of space due to the expanding universe. The result of such measurements (Riess et al., 1998; Perlmutter et al., 1999) implies that the universe is currently in a mildly accelerating phase of expansion.

7.5 NEUTRON STARS

As a core-collapse supernova is initiated, the rapidly rising density in the core results in the electron energy increasing rapidly due to degeneracy effects. The electrons can then bring enough energy to facilitate the endothermic reaction by which electrons combine with protons to make neutrons. This yields a self-gravitating mass of neutrons. Since neutrons are fermions, they also eventually become degenerate once the density is high enough. In a similar manner as for white dwarfs, one can derive a radius-mass relation, which is

$$R = 0.114 \frac{h^2}{G m_n^{8/3}} M^{-1/3}. \tag{7.14}$$

In the above equation, a mass $M = 1.4\,M_\odot$ implies a radius $R = 1.35 \times 10^6$ cm $= 13.5$ km. In that case, the average density of a neutron star would be $\rho = 2.7 \times 10^{14}$ g cm^{-3}, which is comparable to the density of an atomic nucleus. Hence, we can think of a neutron star as a gigantic nucleus held together by gravity!

Neutron stars have very strong gravitation, and the escape speed is

$$v_{\mathrm{esc}} = \left(\frac{2GM}{R}\right)^{1/2} = 1.66 \times 10^{10} \,\mathrm{cm\,s^{-1}} = 0.55\,c\,. \qquad (7.15)$$

An upper mass limit exists for the same reason as for white dwarfs (saturation of degeneracy pressure when $v \sim c$). This upper limit is known as the Tolman–Oppenheimer–Volkoff (TOV) limit. However, its value is uncertain due to an incomplete understanding of the equation of state of dense nuclear matter. Currently, the best estimate of this upper mass limit (Baym et al., 2019) is

$$M_{\mathrm{TOV}} \approx 2.4\,M_{\odot}\,. \qquad (7.16)$$

Indeed, the observational evidence for black holes presented in Section 7.6.2 is consistent with all having masses that exceed the TOV limit.

7.5.1 Pulsars

Pulsars were discovered as very regularly pulsating radio sources by Jocelyn Bell Burnell and Antony Hewish in 1967. By now, about five hundred such sources are known. The characteristic property of a pulsar that differentiates it from other radio sources is its extremely regular period. The periods range from about 1.5 milliseconds to 3 seconds, and the time between pulses is constant to within one part in 10^8. There are very few optical identifications of pulsars, most notably the Crab Nebula Pulsar and the Vela Nebula Pulsar. The Crab Pulsar sits in the center of the Crab Nebula, a very young supernova remnant with an origin during recorded human history. There is evidence that it was observed in many parts of the world. The best known record was made by Chinese astronomers, whose records show that it became visible on 4 July, 1054 AD, could be seen during the daytime for 23 days, and was visible during the night for almost two years.

The Crab pulsar exhibits the same period in gamma rays, X-rays, optical, and radio waves. Soon after their discovery, pulsars were identified with rotating and magnetized neutron stars. But, what is the proof? Well, if the pulse period P corresponds to a rotation period, then breakup can be avoided only if the surface rotation speed v, mass M, and radius R are such that the required centripetal acceleration at the surface is less than or equal to the gravitational acceleration at the surface, i.e.,

$$\frac{v^2}{R} \leq \frac{GM}{R^2}\,. \qquad (7.17)$$

By combining this condition with $P = (2\pi R)/v$ and $M = (4/3)\pi R^3 \rho$, we find a requirement that the average density

$$\rho \geq \frac{3\pi}{G}\frac{1}{P^2}\,. \qquad (7.18)$$

Therefore, if $P = 1$ s, then $\rho \geq 1.4 \times 10^{14}$ g cm^{-3}, which is a typical neutron star density.

7.5.2 Emission Mechanism for Pulsars

In the **lighthouse model** for emission from a pulsar, the magnetic axis is slightly misaligned with the rotation axis. A spinning magnetic field creates an enormous electric field as a consequence of Maxwell's equations of electromagnetism. This pulls some free charged particles off the surface. Note that this means the electron–proton coalescence, which is effective in the interior of the neutron star, is incomplete at the surface. This allows some free charges to exist near the surface and also carry electric currents that generate the magnetic field. Accelerated charged particles that are leaving the neutron star will emit radiation primarily along the poles of the magnetic field. We see this emission as a pulse if/when the beam comes into our line of sight. According to this model, not all neutron stars will appear as pulsars to us, as it depends on a favorable alignment of the magnetic axis with our line of sight.

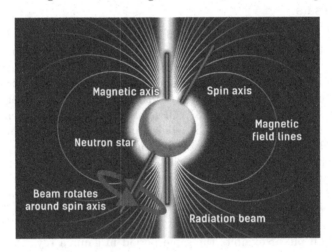

Figure 7.1 An illustration of the lighthouse model of emission from a pulsar, a rapidly rotating neutron star.

The torque from the departing particles also slows down the pulsar's rotation. Estimates of the rate of decrease of rotational energy of the Crab pulsar as its observed period increases by $dP/dt \approx 10^{-5}$ s yr^{-1} show that it approximately equals the rate of emitted energy from the Crab nebula.

7.5.3 Fast Radio Bursts

In 2007, a perplexing new phenomenon was discovered – brief millisecond-duration radio bursts coming from extragalactic sources (Lorimer et al., 2007).

Although pulsars were already known to produce flares of similar duration (giant radio pulses; GRPs), these **fast radio bursts** (FRBs) had to have intrinsic luminosities that were many orders of magnitude greater, given their extragalactic origin. The energetics of the FRBs quickly led to the conclusion that they were associated with **magnetars**, a class of neutron stars with extreme surface magnetic fields of strength $\sim 10^{14}$ G. According to this hypothesis, the strain induced by an intense magnetic field could lead to violent rearrangements of the neutron star crust in a **starquake** event. The released energy could be transported through the **magnetosphere** and released as gamma rays, X-rays, and radio waves (FRBs). This hypothesis required that the known magnetars in our Galaxy should also undergo such events, even though their previously detected GRPs were many orders of magnitude fainter.

In April 2020, the *Canadian Hydrogen Intensity Mapping Experiment* (CHIME) and the *Survey for Transient Astronomical Radio Emission 2* (STARE2) experiment in the USA detected an FRB from the direction and distance of the known Galactic magnetar SGRJ1935+2154 (Bochenek et al., 2020). Gamma ray and X-ray emission were also detected by several satellites. This observation has proven that magnetars can indeed produce FRBs that are bright enough to be observed when the source is extragalactic. Since most radio telescopes have a narrow field-of-view, the extrapolation from previous FRB detections (only about two dozen were detected between 2007 and mid-2017) was that thousands of FRBs reach Earth every day. CHIME has a unique design with an unusually large field-of-view for a radio telescope and is greatly expanding the number of known extragalactic FRBs. Indeed, CHIME has cataloged over 1000 FRBs as of mid-2020. Perhaps, the study of FRBs can lead to new physical insights into the structure of neutron stars, which are some of the most exotic objects in the universe. Alternately, their detection from great distances can serve as a cosmological probe of the large-scale structure of the universe.

7.6 BLACK HOLES

Finally, we arrive at the fascinating objects known as black holes, which are formed when an object collapses indefinitely under its own weight. This bizarre phenomenon represents the ultimate victory of gravity, forming a metaphorical hole in space-time. Stellar remnants of mass greater than the TOV limit of $\approx 2.4\,M_\odot$ go into indefinite collapse toward a singular point of zero volume and infinite density, popularly called a singularity. However, there exists an important length scale where the escape speed is

$$v_{\text{esc}} = \left(\frac{2GM}{R} \right)^{1/2} = c \,. \tag{7.19}$$

This scale is

$$R = R_{\text{Sch}} \equiv \frac{2GM}{c^2} \,, \tag{7.20}$$

known as the Schwarzschild radius. It is named after Karl Schwarzschild (1873 - 1916), who derived the relation from the full equations of the theory of general relativity developed by Albert Einstein. His was the first exact solution of the equations of general relativity, and was derived in the same year, 1915, that the theory was introduced. If an object is compressed within R_{Sch}, then nothing, neither matter nor radiation, can escape the pull of gravity of the object. This is how we define a black hole.

7.6.1 Physical Meaning of the Schwarzschild Radius

The **event horizon** is the effective surface of a black hole, even if it is not a material boundary. It is identified with the radius R_{Sch}, and means that no information can be obtained from inside it. Since there is no actual boundary at the event horizon, a traveler making the journey to the center of a black hole would pass straight through. However, strong tidal forces would rip the traveler apart, since the Roche limit is typically much larger than the Schwarzschild radius.

Is there actually a singularity at the center of a black hole? Although we can never observe it, this is a fascinating topic that tests the limits of known physics. While our current theories of gravity imply that gravity wins indefinitely and the collapsing stellar remnant collapses all the way to a singularity, we also know that quantum mechanics does not allow an indefinitely small localization of matter. The two ideas will need to be reconciled in a full theory of **quantum gravity**, which is yet to be developed.

7.6.2 Observational Signatures

The radiation emerging from matter as it falls in toward the event horizon can be estimated with some simple physical principles and a little bit of insight. Any mass element dm that falls into the event horizon in a time dt, originating at a large distance $R \gg R_{\text{Sch}}$, will release a gravitational energy per unit mass GM/R_{Sch}. If most of this energy is ultimately converted to thermal and then radiant energy in a swirling accretion disk near the event horizon, then we can estimate the resulting luminosity as

$$L \approx \left| \frac{dE_{\text{grav}}}{dt} \right| \approx \left(\frac{GM}{R_{\text{Sch}}} \right) \frac{dm}{dt} . \tag{7.21}$$

Using the definition of R_{Sch}, we can rewrite this as

$$L \approx \frac{1}{2} \frac{dm}{dt} c^2, \tag{7.22}$$

which means that a significant portion of the incoming mass-energy is extracted. The fraction of incoming rest mass energy that can be released in the form of radiation is 0.5 in the above equation, far exceeding the fraction 0.007 that is liberated in the fusion of hydrogen to helium. Ultimately, gravity power exceeds nuclear power if matter can fall in to a region of order the Schwarzschild radius. This model can explain the extremely high luminosities from some binary systems, e.g., Cygnus X-1. We observe X-ray emission from material falling onto the black hole that is heated to $\sim 10^6$ K temperatures and is radiating as it is about to enter the event horizon.

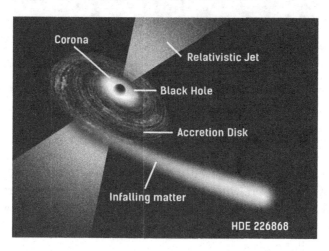

Figure 7.2 An illustration of mass transfer in a binary system, with mass from a star being captured by a black hole companion. The swirling mass of gas outside the event horizon effectively transforms the released gravitational energy of the infalling gases into radiant luminosity (mostly in the X-ray band) that can be seen by astronomers. The first such observed X-ray binary is the system Cygnus X-1, also known as HDE 226868.

The black holes in binary X-ray systems are identifiable since they emit negligible radiation in visible light, while at the same time, an analysis of the binary motions implies that their mass exceeds the TOV limit. Theoretical models of stellar evolution generally predict that an isolated massive Pop I star will suffer enough mass loss during its late stage of evolution such that a remnant black hole should have a mass no more than $\sim 10\,M_\odot$. The black hole candidates in X-ray binary systems are indeed found to have a range of masses that are above 3 $M_\odot$ and below 20 $M_\odot$.

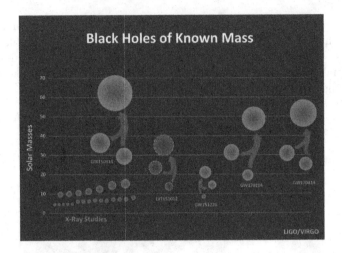

Figure 7.3 A chart showing the masses of binary pairs of objects that underwent merger and whose resulting gravitational waves were detected by the LIGO and Virgo detectors from 2015 to 2017. Also shown are the masses of black holes detected in X-ray binary systems. Credit: LIGO Scientific Collaboration.

An interesting challenge to these ideas of stellar evolution comes from the recent observations of the merger of binary black hole pairs by the *Laser Interferometric Gravitational-Wave Observatory* (LIGO). The first detection of a binary black hole merger, labeled GW150914 (Abbott et al., 2016), was inferred by analysis of the gravitational wave signal to correspond to two black holes of mass 36 $M_\odot$ and 29 $M_\odot$, respectively. Subsequent detections until 2017 extended the range from 7.5 $M_\odot$ to 50 $M_\odot$, as seen in Figure 7.3. By 2020, the upper mass limit of detected merging black holes has been pushed to about 80 $M_\odot$. Clearly, we have much to learn about the formation of black holes and the origin of their masses.

Bibliography

Abbott, B. P., Abbott, R., Abbott, T. D., Abernathy, M. R., Acernese, F., et al. (2016). Observation of gravitational waves from a binary black hole merger. *Phys. Rev. Lett.*, *116*, 061102.

Adams, F. C., Lada, C. J., & Shu, F. H. (1987). Spectral evolution of Young stellar objects. *ApJ*, *312*, 788.

Alves, J. F., Lada, C. J., & Lada, E. A. (2001). Internal structure of a cold dark molecular cloud inferred from the extinction of background starlight. *Nature*, *409*(6817), 159–161.

Andre, P., Ward-Thompson, D., & Barsony, M. (1993). Submillimeter Continuum observations of rho Ophiuchi A: The Candidate Protostar VLA 1623 and Prestellar Clumps. *ApJ*, *406*, 122.

Andrews, S. M., Huang, J., Pérez, L. M., Isella, A., Dullemond, C. P., et al. (2018). The disk substructures at high angular resolution project (DSHARP). I. Motivation, Sample, Calibration, and Overview. *ApJ*, *869*(2), L41.

Auddy, S., Basu, S., & Valluri, S. R. (2016). Analytic models of brown dwarfs and the substellar mass limit. *Advances in Astronomy*, *2016*, 574327.

Basu, S. (2012). Brown-dwarf origins. *Science*, *337*(6090), 43.

Basu, S., Gil, M., & Auddy, S. (2015). The MLP distribution: A modified lognormal power-law model for the stellar initial mass function. *MNRAS*, *449*(3), 2413–2420.

Basu, S., Johnstone, D., & Martin, P. G. (1999). Dynamical evolution and ionization structure of an expanding superbubble: Application to W4. *ApJ*, *516*(2), 843–862.

Basu, S., Mouschovias, T. C., & Paleologou, E. V. (1997). Dynamical effects of the Parker instability in the interstellar medium. *ApJ*, *480*(1), L55–L58.

Baym, G., Furusawa, S., Hatsuda, T., Kojo, T., & Togashi, H. (2019). New neutron star equation of state with quark-hadron crossover. *ApJ*, *885*(1), 42.

Bochenek, C. D., Ravi, V., Belov, K. V., Hallinan, G., Kocz, J., Kulkarni, S. R., & McKenna, D. L. (2020). A fast radio burst associated with a Galactic magnetar. *Nature*, *587*(7832), 59–62.

Bodenheimer, P., & Sweigart, A. (1968). Dynamic collapse of the isothermal sphere. *ApJ*, *152*, 515.

Bryson, S., Kunimoto, M., Kopparapu, R. K., Coughlin, J. L., Borucki, W. J., et al. (2021). The occurrence of rocky habitable-zone planets around solar-like stars from Kepler data. *AJ*, *161*(1), 36.

Cecil, G., Bland-Hawthorn, J., Veilleux, S., & Filippenko, A. V. (2001). Jet- and wind-driven ionized outflows in the superbubble and star-forming disk of NGC 3079. *ApJ*, *555*(1), 338–355.

Chabrier, G., & Baraffe, I. (2000). Theory of low-mass stars and substellar objects. *ARA&A*, *38*, 337–377.

Chandrasekhar, S., & Fermi, E. (1953). Magnetic fields in spiral arms. *ApJ*, *118*, 113.

Davis, J., Leverett, & Greenstein, J. L. (1951). The polarization of starlight by aligned dust grains. *ApJ*, *114*, 206.

Davis, L. (1951). The strength of interstellar magnetic fields. *Physical Review*, *81*(5), 890–891.

Dickey, J. M., & Lockman, F. J. (1990). H I in the Galaxy. *ARA&A*, *28*, 215–261.

Duchêne, G., & Kraus, A. (2013). Stellar multiplicity. *ARA&A*, *51*(1), 269–310.

Ewen, H. I., & Purcell, E. M. (1951). Observation of a line in the galactic radio spectrum: Radiation from galactic hydrogen at 1,420 Mc./sec. *Nature*, *168*, 356.

Field, G. B., Goldsmith, D. W., & Habing, H. J. (1969). Cosmic-ray heating of the interstellar gas. *ApJ*, *155*, L149.

Hall, J. S. (1949). Observations of the polarized light from stars. *Science*, *109*(2825), 166–167.

Hartmann, J. (1904). Investigations on the spectrum and orbit of delta Orionis. *ApJ*, *19*, 268–286.

Haslam, C. G. T., Salter, C. J., Stoffel, H., & Wilson, W. E. (1982). A 408 MHz all-sky continuum survey. II. The atlas of contour maps. *A&AS*, *47*, 1–143.

Hayashi, C. (1961). Stellar evolution in early phases of gravitational contraction. *PASJ*, *13*, 450–452.

Hayashi, C. (1966). Evolution of protostars. *ARA&A*, *4*, 171.

Hayashi, C., & Nakano, T. (1963). Evolution of stars of small masses in the pre-main-sequence stages. *Progress of Theoretical Physics*, *30*(4), 460–474.

Hiltner, W. A. (1949). Polarization of light from distant stars by interstellar medium. *Science*, *109*(2825), 165.

Iben, J., Icko (1967). Stellar evolution within and off the main sequence. *ARA&A*, *5*, 571.

Jeans, J. H. (1902). The stability of a spherical nebula. *Philosophical Transactions of the Royal Society of London. Series A, Containing Papers of a Mathematical or Physical Character*, *199*, 1–53.

Johnson, H. L., Mitchell, R. I., & Iriarte, B. (1962). The color-magnitude diagram of the hyades cluster. *ApJ*, *136*, 75.

Kompaneets, A. S. (1960). A point explosion in an inhomogeneous atmosphere. *Soviet Physics Doklady*, *5*, 46.

Kumar, S. S. (1963). The structure of stars of very low mass. *ApJ*, *137*, 1121.

Larson, R. B. (1969). Numerical calculations of the dynamics of collapsing proto-star. *MNRAS*, *145*, 271.

Lin, D. N. C., Bodenheimer, P., & Richardson, D. C. (1996). Orbital migration of the planetary companion of 51 Pegasi to its present location. *Nature*, *380*(6575), 606–607.

Lorimer, D. R., Bailes, M., McLaughlin, M. A., Narkevic, D. J., & Crawford, F. (2007). A bright millisecond radio burst of extragalactic origin. *Science*, *318*(5851), 777.

Machida, M. N., Inutsuka, S.-i., & Matsumoto, T. (2007). Magnetic fields and rotations of protostars. *ApJ*, *670*(2), 1198–1213.

Machida, M. N., Inutsuka, S.-i., & Matsumoto, T. (2008). High- and low-velocity magnetized outflows in the star formation process in a gravitationally collapsing cloud. *ApJ*, *676*(2), 1088–1108.

Masuda, K. (2014). Very low density planets around Kepler-51 revealed with transit timing variations and an anomaly similar to a planet-planet eclipse event. *ApJ*, *783*(1), 53.

Masunaga, H., & Inutsuka, S.-i. (2000). A radiation hydrodynamic model for protostellar collapse. II. The second collapse and the birth of a protostar. *ApJ*, *531*(1), 350–365.

Mathewson, D. S., & Ford, V. L. (1970). Polarization observations of 1800 stars. *MmRAS*, *74*, 139.

Mayor, M., & Queloz, D. (1995). A Jupiter-mass companion to a solar-type star. *Nature*, *378*(6555), 355–359.

Mouschovias, T. C., & Spitzer, L., Jr. (1976). Note on the collapse of magnetic interstellar clouds. *ApJ*, *210*, 326.

Muench, A. A., Lada, E. A., Lada, C. J., & Alves, J. (2002). The Luminosity and Mass Function of the Trapezium Cluster: From B Stars to the Deuterium-burning Limit. *ApJ*, *573*(1), 366–393.

Nefs, S. V., Birkby, J. L., Snellen, I. A. G., Hodgkin, S. T., Sipőcz, B. M., et al. (2013). A highly unequal-mass eclipsing M-dwarf binary in the WFCAM transit survey. *MNRAS*, *431*(4), 3240–3257.

Onishi, T., Mizuno, A., Kawamura, A., Ogawa, H., & Fukui, Y. (1996). A C 18O Survey of dense cloud cores in Taurus: Core properties. *ApJ*, *465*, 815.

Paladini, R., Davies, R. D., & De Zotti, G. (2004). Spatial distribution of Galactic HII regions. *MNRAS*, *347*(1), 237–245.

Parker, E. N. (1966). The Dynamical state of the interstellar gas and field. *ApJ*, *145*, 811.

Penston, M. V. (1969). Dynamics of self-gravitating gaseous spheres-III. Analytical results in the free-fall of isothermal cases. *MNRAS*, *144*, 425.

Perlmutter, S., Aldering, G., Goldhaber, G., Knop, R. A., Nugent, P., et al. (1999). Measurements of Ω and Λ from 42 High-Redshift Supernovae. *ApJ*, *517*(2), 565–586.

Planck Collaboration, Ade, P. A. R., Aghanim, N., Alves, M. I. R., Arnaud, M., Arzoumanian, D., et al. (2016). Planck intermediate results. XXXV. Probing the role of the magnetic field in the formation of structure in molecular clouds. *A&A*, *586*, A138.

Plaskett, J. S., & Pearce, J. A. (1930). The motions and distribution of inter-stellar matter. *MNRAS*, *90*, 243–268.

Raghavan, D., McAlister, H. A., Henry, T. J., Latham, D. W., Marcy, G. W., et al. (2010). A survey of stellar families: Multiplicity of solar-type stars. *ApJS*, *190*(1), 1–42.

Riess, A. G., Filippenko, A. V., Challis, P., Clocchiatti, A., Diercks, A., et al. (1998). Observational evidence from supernovae for an accelerating universe and a cosmological constant. *AJ*, *116*(3), 1009–1038.

Safronov, V. S. (1972). *Evolution of the protoplanetary cloud and formation of the earth and planets.*

Saha, M. N. (1921). On a Physical Theory of Stellar Spectra. *Proceedings of the Royal Society of London Series A*, *99*(697), 135–153.

Shu, F. H. (1982). *The Physical Universe.* University Science Books.

Silva, D. R., & Cornell, M. E. (1992). A new library of stellar optical spectra. *ApJS*, *81*, 865.

Tassis, K., Ramaprakash, A. N., Readhead, A. C. S., Potter, S. B., Wehus, I. K., et al. (2018). PASIPHAE: A high-Galactic-latitude, high-accuracy optopolarimetric survey. *arXiv e-prints*, (arXiv:1810.05652).

Taylor, J. H., & Cordes, J. M. (1993). Pulsar distances and the galactic distribution of free electrons. *ApJ*, *411*, 674.

Toomre, A. (1964). On the gravitational stability of a disk of stars. *ApJ*, *139*, 1217–1238.

Townsley, L. K., Broos, P. S., Garmire, G. P., Bouwman, J., Povich, M. S., Feigelson, E. D., Getman, K. V., & Kuhn, M. A. (2014). The massive star-forming regions omnibus X-Ray catalog. *ApJS*, *213*(1), 1.

Trumpler, R. J. (1930). Absorption of light in the galactic system. *PASP*, *42*(248), 214.

Vorobyov, E. I., & Basu, S. (2006). The burst mode of protostellar accretion. *ApJ*, *650*(2), 956–969.

Vorobyov, E. I., & Basu, S. (2010). The burst mode of accretion and disk fragmentation in the early embedded stages of star formation. *ApJ*, *719*(2), 1896–1911.

Wang, J., & Zhong, Z. (2018). Revisiting the mass-luminosity relation with an effective temperature modifier. *A&A*, *619*, L1.

Index

Printed in the United States
by Baker & Taylor Publisher Services